I0084588

Internet Diplomacy

Digital Technologies and Global Politics

Series Editors:
Andrea Calderaro and Madeline Carr

While other disciplines like law, sociology, and computer science have engaged closely with the Information Age, international relations scholars have yet to bring the full analytic power of their discipline to developing our understanding of what new digital technologies mean for concepts like war, peace, security, cooperation, human rights, equity, and power. This series brings together the latest research from international relations scholars—particularly those working across disciplines—to challenge and extend our understanding of world politics in the Information Age.

Governing Cyberspace: Behaviour, Power and Diplomacy
Edited by Dennis Broeders and Bibi van den Berg

Internet Diplomacy: Shaping the Global Politics of Cyberspace
Edited by Meryem Marzouki & Andrea Calderaro

Internet Diplomacy

Shaping the Global Politics of Cyberspace

Edited by Meryem Marzouki
and Andrea Calderaro

ROWMAN & LITTLEFIELD
Lanham • Boulder • New York • London

Published by Rowman & Littlefield
An imprint of The Rowman & Littlefield Publishing Group, Inc.
4501 Forbes Boulevard, Suite 200, Lanham, Maryland 20706
www.rowman.com

86-90 Paul Street, London EC2A 4NE

Copyright © 2022 by The Rowman & Littlefield Publishing Group, Inc.

All rights reserved. No part of this book may be reproduced in any form or by any electronic or mechanical means, including information storage and retrieval systems, without written permission from the publisher, except by a reviewer who may quote passages in a review.

British Library Cataloguing in Publication Information Available

Library of Congress Cataloging-in-Publication Data
Names: Marzouki, Meryem, editor. | Calderaro, Andrea, editor.
Title: Internet diplomacy : shaping the global politics of cyberspace / edited by Meryem Marzouki and Andrea Calderaro.
Description: Lanham, Maryland : Rowman & Littlefield, an imprint of The Rowman & Littlefield Publishing Group, Inc., [2022] | Series: Digital technologies and global politics | Includes bibliographical references and index. | Summary: "This book offers an approach to internet diplomacy that extends to foreign affairs and international relations, and with regard to all emerging international tensions clustered around digital environments, including cybersecurity and internet governance"— Provided by publisher.
Identifiers: LCCN 2022001241 (print) | LCCN 2022001242 (ebook) | ISBN 9781538168165 (paper) | ISBN 9781538161180 (epub)
Subjects: LCSH: Internet governance. | Internet—Political aspects.
Classification: LCC TK5105.8854 .I57 2022 (print) | LCC TK5105.8854 (ebook) | DDC 384.3/34—dc23/eng/20220328
LC record available at https://lccn.loc.gov/2022001241
LC ebook record available at https://lccn.loc.gov/2022001242

Contents

1 Global Internet Governance: An Uncharted Diplomacy Terrain?   1
Meryem Marzouki and Andrea Calderaro

PART I: INTERNET GOVERNANCE AS A DIPLOMACY ISSUE

2 Undiplomatic Ties: When Internet Blocks Intermediation   21
Yves Schemeil

3 Diplomacy and Internet Governance: A Conceptual
Re-assessment   45
Katharina E. Höne

4 Discourse Coalitions in Internet Governance: Shaping Global
Policy by Narratives and Definitions   61
Mauro Santaniello and Nicola Palladino

**PART II: INTERNET GOVERNANCE AS A SCIENCE
DIPLOMACY AREA**

5 Science Diplomacy and Internet Governance:
Opportunities and Pitfalls   87
Robin Mansell

6 Crafting Science Diplomacy in Comparative Perspective:
The Case of US Internet Governance   107
Nanette S. Levinson

7 Modes of Internet Governance as Science Diplomacy:
What Might the EU Learn from the US Cybersecurity Policy?   119
Francesco Amoretti and Domenico Fracchiolla

**PART III: CASE STUDIES OF INTERNET GOVERNANCE
DIPLOMACY**

8 Free Trade Governance and Data Flow: The TiSA's Agreement
   Negotiation as a Case for Governmentality Studies   143
   Maria Francesca De Tullio and Giuseppe Micciarelli

9 National Sovereignty, Global Policy, and the Liberalization of
   Telecommunications Markets   161
   Claire Peters

10 In This Bright Future You Can't Forget Your Past:
    Debating the 'Right to Be Forgotten' in Latin America   177
    Jean-Marie Chenou

11 Policy Diffusion and Internet Governance:
    Reflections on Copyright and Privacy   195
    Krisztina Rozgonyi and Katharine Sarikakis

List of Acronyms   211

References   217

Index   255

About the Editors and Contributors   267

Chapter One

Global Internet Governance

An Uncharted Diplomacy Terrain?

Meryem Marzouki and Andrea Calderaro

On Friday, 27 January 2017, then Danish Foreign Minister Anders Samuelsen announced his plan for Denmark to become the first nation ever to appoint a 'digital ambassador' (Jarlner and Koch 2017). Noting that giant US digital companies – such as the so-called GAFA (Google, Apple, Facebook and Amazon) – 'affect Denmark just as much as entire countries', he concluded that 'these companies have become a type of new nations and [Denmark] need[s] to confront that'. With the cautious mention that '[Denmark] will, of course, maintain [its] old way of thinking in which [it] foster[s] [its] relationships with other countries. But [Denmark] simply need[s] to have closer ties to some of the companies that affect [it]', the minister claimed: 'We are sending a signal that includes that royal crown and our entire diplomacy' (Jarlner and Koch 2017).

This announcement was, at the same time, a pioneering move and an emblematic sign of the increased centrality of diplomacy to ease tensions around the transnational governance of digital challenges in foreign relations. However, given the peculiarity of the digital domain, characterized by the transnational nature of its infrastructure, the variety of actors beyond those of the state necessary to negotiate the technical, policy, economic, and security aspects of the internet (DeNardis 2018[1]), the translation of this new dimension of diplomacy into a tangible concept and practice is still limited and vague.

Digital issues generate new spaces of conflicts as such, where new diplomatic practices take space to facilitate the negotiations among parties about the governance, policy developments and technical solutions of the internet (Calderaro and Kavada 2013; Radu 2019). As a result of this, in order to understand the shift of diplomatic practices engaging with the digital domain, we need to expand our understanding of the global governance of the internet and identify the tools, venues, and processes, so as to adequately address this new dimension of diplomacy (Broeders and van den Berg 2020b).

While digital diplomacy is traditionally referred to as the use of digital technologies to engage in diplomatic dialogues (Bjola and Zaiotti, 2021), the adoption of diplomatic practices to negotiate the variety of challenges related to the functioning of the digital domain is often framed as cyber diplomacy (Riordan 2019) which is defined as 'the use of diplomatic resources and the performance of diplomatic functions to secure national interests with regards to the cyberspace' (Barrinha and Renard 2017, 3). Given the focus on the role played by diplomats in ensuring the security of 'national interest', this definition mostly addresses practices performed by traditional diplomats representing the interest of state actors (Thomson 1995). However, it is crucial to take into account that state actors are only partially responsible for the functioning of the internet. If state actors do have the responsibility to provide the legal and policy framework facilitating citizen's online experience, industry traditionally leads the development of the internet infrastructure and most of its services (Powers and Jablonski 2015; Carr 2016). We argue that the concept of diplomacy in the global governance of the internet goes beyond traditional diplomatic practices solely performed by states' representatives. The stability, safety, and economy of the internet involve a variety of actors and expertise beyond state actors in the negotiations of technical protocols, international agreements, legislations, and forms of governance accountability in a new dimension of diplomacy (Calderaro 2021), that, with this book, we address as Internet Diplomacy.

Our approach to Internet Diplomacy extends to foreign affairs and international relations, and with regard to all emerging international tensions clustered around digital environments, including cybersecurity, internet governance, and the political economy of the internet. In other words, we refer to Internet Diplomacy as the adoption of diplomatic practices by both state and non-state actors, to negotiate any technical, legal, policy, economy, security issues, and practices related to the functioning of the internet.

THE MULTISTAKEHOLDERISM OF INTERNET DIPLOMACY

Given the increasing centrality of the governance of the digital field in global politics, we are witnessing a growing need to better understand recent transformations of international diplomacy in this context, their drivers and their nature, whether and how they might change European and transnational power relations and, ultimately, which values they carry and channel on the global scene. Since the UN World Summit on Information Society (WSIS[2]) in 2003–2005, the United Nations has established that internet governance processes should be institutionalized in an open and inclusive manner through multistakeholder participation. This decision has extended the invitation at the table of diplomatic negotiations to actors beyond governments, including the business sector and civil society.

Since then, multistakeholder participation along this line has characterized several initiatives by the UN and its agencies in the internet governance field, as well as many thematic or regional international organizations (Levinson and Marzouki 2015). As part of recent moves, in July 2018, the United Nations secretary general appointed 'a High-Level Panel on Digital Cooperation', co-chaired by Melinda Gates, Co-Chair of the Bill & Melinda Gates Foundation, and Jack Ma, then Executive Chairman of Alibaba Group (UN 2018). In December 2018, the United Nations General Assembly (UNGA) adopted a multistakeholder approach by launching the Open-Ended Working Group (OEWG) to complement the process led by the Group of Governmental Experts (UN GGE[3]) on 'advancing responsible State behaviour in cyberspace in the context of international security'. While the UN GGE only included diplomats representing 25 selected UN member states, the UN OEWG[4] in the 'field of information and telecommunications in the context of international security' welcomed all UN member states and representatives of civil society and the business sector to 'act on a consensus basis' to contribute to 'further develop the rules, norms and principles of responsible behavior of States listed [in the same resolution], and the ways for their implementation' (UNGA 2018). These UN initiatives and their outcomes offer additional evidence on how a multistakeholder approach adopted for the negotiation of internet related issues has set the standards for present and future development of Internet Diplomacy. All this has an impact on the broader field of global governance studies and has set important mutations in diplomatic practices beyond the cyber dimension (Scholte 2005).

Given also the variety of actors and experts engaged in the transnational governance of the digital domain, the definition of commonly shared solutions among stakeholders is particularly challenged by the fast-moving target of the negotiations. In particular, the capacity of stakeholders to make informed choices and agree on issues that go beyond the engineering dimension of the internet is challenged by the rapid developments of the technical, social, and market aspects of the internet. In other words, traditionally, technology evolves quicker than the capacity of policy makers to understand the implications of the technological shift, which is a major obstacle for diplomats and their role in negotiations taking place around digital policy making. As a result of this, technological developments regularly challenge national, regional, and global policy making, in terms of sovereignty and other political, legal, economic, social, cultural, and societal choices. The mutation of diplomatic practices in Internet Diplomacy reflects these challenges.

AT THE CROSSROADS OF GLOBALIZATION AND DIGITALIZATION

Diplomacy in the transnational governance of the digital domain is called upon to tackle the deep and multifold mutation resulting from multifaceted digital disruptions that affect every aspect of current social, economic, and political life. Together with opportunities, these transformations generate challenges in terms of sovereignty, economic development, social cohesion, political and cultural values, and legal and policy frameworks (DeNardis 2018). The cross-field nature of the impact of the internet imposes diplomats to adopt a multidisciplinary understanding of the issues, by combining their representative role with scientific advice. Contrary to other fields of global governance clustered around the role of states and international organizations (Zürn 2018), in global internet governance scientific and technical expertise join forces with political influence and diplomatic action to address digital challenges (Kaltofen and Acuto 2018), channel democratic values, and share knowledge to build common visions (Scholte 2005).

Moreover, given the transnational nature of the internet, the implementation of digital policies at the national level might generate impacts on a global scale, with potential amplified consequences on global politics. For this reason, we are witnessing an increasing need to enhance international cooperation beyond national borders and national legislations in order to define a consistent and inclusive transnational governance approach to the cyber domain (Calderaro and Craig 2020).

By global internet governance, we intend not only the restricted issue of managing internet technical resources (infrastructure, protocols, and domain names) and technical standards setting (Harcourt, Christou, and Simpson 2020), but, as defined by the Working Group on Internet Governance (WGIG 2005) and adopted in 2005 by the United Nations at WSIS, an extensive set of issues ranging from the 'administration of the root zone files and system' to 'capacity building' and the 'meaningful participation in global policy development' as well as a whole set of human rights and consumer rights issues directly at stake in the governance of information and communication processes (Brousseau and Marzouki 2012).

Among the cross-cutting challenges that domestic and foreign policies are facing, we can identify the following (DeNardis 2018): The difficulty to keep pace with numerous internet innovations in order to make informed choices and decisions on issues that may appear only technical; the difficulty to understand and conciliate roles and positions of a great variety of actors and stakeholders; the need to identify the different exchanges and dialogues in regional and international fora and to navigate in these waters; and the requirement to be aware of their evolving strategies of transaction and coalition, while internet governance has proven to be much more than a public policy issue in light of the essential characteristics of this network (the interconnection is global; its management is distributed) and as it has historically been privately coordinated and operated. Finally, there is a need to channel and maintain democratic values in global internet governance processes, namely that of sustainability, participatory governance, openness and transparency in policies and markets, human rights, social justice, and social cohesion, as well as democracy and the rule of law. As recent discussions on European digital sovereignty are showing (Christakis 2020; Madiega 2020), this challenge is particularly difficult to address with private US firms, such as the GAFA and other internet giants, including the emerging role played by Chinese ones, dominating the innovation market.

This book stems from the observation of a number of mutations at the crossroads of globalization and digitalization. These mutations concern both the global governance of the online world, which has been facing several disruptions, and diplomacy itself, which has been experiencing important transformations. In our view, both categories of mutations must be addressed at the same time, analyzing and understanding the digital disruptive trends they create while exploring how Internet Diplomacy could be an effective mean to address such disruptive trends to keep channeling democratic values in global internet governance processes. Focusing on global internet governance allows exploring, in a consistent way, almost all issues related to globalization and, therefore, at the heart of foreign policy international relations discussions: sovereignty, security, trade, finance and taxation, economic and social transformations (including labor) as well as human rights, democracy and the rule of law. These particularly wide implications of global internet governance result, obviously, from the fact that the internet has become an integral part of the political, economic and social life in all their dimensions and from the fact that the network raises, by construction, cross-border issues.

UNDERSTANDING MULTIFACETED DIGITAL DISRUPTIONS

Since the internet reached a wide public in the mid-90s, technical, business, marketing, communication, and innovations have profoundly transformed power relations, in social, economic, normative, institutional, and geopolitical terms. Earlier controversies regarding human rights implications of internet use emerged at the national level, leading policy makers to address issues related to conflicts of rights and conflicts of jurisdictions[5]. Becoming even more prominent with the development of social media platforms and their centrality in society's online experience, these debates also led to procedural issues related to the role and liability of internet intermediaries and algorithms, in terms of legal, technical and economic aspects.

Ten years later, the United Nations held the WSIS with the ambition to define a globally recognized governance model for the internet, and how internet governance processes should be institutionalized in an open and inclusive manner through multistakeholder participation. One of the main outcomes of the Summit was the creation of the Internet Governance Forum (IGF[6]), which, since 2006, has served as the main open forum for such dialogue among all interested stakeholders.

The launch of these initiatives has de facto institutionalized international cooperation in the digital domain. Consequently, all global actors, including the EU and its member states, have started engaging in these new global processes in accordance with their own foundational values and national priorities (Mueller 2010). At the same time, they had to face the transnational nature of the internet and, more specifically, the extraterritorial effect and other kinds of deliberate or serendipitous externalities of internet-related public or private policies and actions. Since this early stage, the global governance of the internet is still clustered around a series of contentious topics that have not yet found long-term solutions. Data protection, data trade, core functions of the internet architecture, intellectual property rights, electronic surveillance, net neutrality, human rights, and the digital divide have traditionally characterized negotiations in the domain of internet governance. Following decades of generalized optimism on the beneficial impact of the internet on society, politics, and economy, concerns emerged also on the potential threats of the internet, and, in recent years, the major emphasis has been on cybersecurity, artificial intelligence, algorithmic governance, platform regulation, and digital sovereignty. We classify these issues and the disruptions they create into four main categories:

Governance disruptions include all the transformations related to state and non-state agents involved in internet governance, including who they are, what is their nature and relevance, what are their strategies, how they coalesce or divide, what kind of relationships they establish among them, how they proceed to advance their views, and, ultimately, how power relations are transforming globally in political and institutional terms. Such agents include state and non-state (civil society, business sector) actors, as well as more inconspicuous actors in this field, such as technical or other epistemic communities, courts, parliaments, regulatory agencies, and intergovernmental organizations. Not only human, organizational, and institutional actors, but also artifactual agents need to be studied and understood, such as internet governance processes and instruments, including architectures, protocols, and algorithms.

Democracy disruptions relate to the transformation of national and international hard and soft law, regulation and private practices, and how they may impact, in the online environment: the substance of human rights and their balance in a democratic society; the rule of law principle; and the democratic values of legitimacy, transparency, accountability, participation, and fairness. This involves procedural as well as substantive transformations and includes, inter alia, the transfer of some states' responsibilities and prerogatives to private intermediaries, the practice of profiling (by private and public actors for marketing and security purposes, respectively) and the datafication trends.

Economy disruptions, where we include all transformations pertaining to trade, finance, and taxation, economy, as well as labor and working conditions, mostly relate to the challenges and opportunities of the 'Uberisation' or so-called sharing economy and its many business and marketing models (Marsden 2015). They also relate to the outsourcing of certain (not necessarily digital, such as after-sale services and various hotlines) economic activities when this is made possible by digitalization.

Geopolitics disruptions relate to the transformations of the global digital geopolitical order. Issues such as network neutrality and the so-called fragmentation of the internet need to be addressed here as a dialectic movement between globalization and renationalization of the digitized world. In addition, transformations in international development and international development aid (especially the emergence of private initiatives, alone or in coordination with state actors) are also part of this category: global internet access, zero-rating policies, their impact on public policies, and the transformed kinds of digital divides they may lead to need also to be studied from the point of view of the new (digital) world (dis)orders they may create. The multiple dimensions of cybersecurity also fall into the category of geopolitical disruptions, as do the provision, control, or prohibition of specific technologies and equipment, such as those used for surveillance, including biometrics.

Many aspects of these four categories of disruptions are the subjects of academic research, as part of profound mutations carrying important implications far beyond the sole online domain: platformization and datafication of the economic and social life, from social networks to the so-called sharing economy; multistakeholderisation of the institutional governance processes, where various categories of stakeholders are involved, together with nation states, in policy arenas and decision-making related to several diplomatic issues; renormativisation, or the reorganization and the reformulation of normative frameworks, following the increasing role of private actors; and fragmentation, the process by which access to the global internet becomes subject to barriers to entry, whether such barriers are of a technical, legal, economic, or social nature and whether their purpose is to discriminate access to infrastructure, software, applications, services, or usages. However, analyzing governance, democracy, economy, and geopolitics mutations to fully understand them in their systemic nature and to unfold their consequences requires a highly multidisciplinary approach, that we identify as the first component of an Internet Diplomacy research agenda.

THE CHANGING NATURE OF DIPLOMACY

These digital disruptions and their effects resonate with diplomatic transformations, as observed in the dedicated literature. Contemporary diplomacy has itself been facing multiple challenges due to the combination of globalization and, more recently, digitalization (Hocking and Melissen 2015). As analyzed by Pouliot and Cornut (2015), the very nature of diplomacy is changing, along different characteristics:

Diplomacy agents: diplomats are no longer only official governmental agents, but also different categories of state and non-state actors interacting with foreign affairs officials as well as among themselves. In this respect, Wiseman argues (2004) that polylateralism constitutes a third form or dimension of diplomacy, in addition to bilateralism and multilateralism. The internet governance world prefers the related concept of multistakeholderism (ITU 2013), coined in management circles, most notably the Davos World Economic Forum (WEF), with the stakeholder theory having been developed by Klaus Schwab in 1971 at the WEF foundation. The concept also found its way to the UN and its agencies and bodies, where it is centered on states as central agents, rather than on corporations, as initially envisioned by Schwab (Gleckman 2012). Multistakeholderism has now widely spread, as both a concept and a mechanism, in internet governance as well as in almost all global governance fields (Scholte 2020).

Diplomacy fields: diplomacy has developed over time, from strict foreign affairs negotiations as an alternative to war, into a myriad of formal and informal discussions on almost any matter, especially with globalization. While the initial raison d'état is often still the ultimate objective of diplomacy, it is no longer restricted to the security field and now encompasses many other components for the stability of a nation and the welfare of its citizens, including economic and social wealth, access to critical resources, and access to knowledge (Cooper, Heine, and Thakur 2013a). Moreover, in addition to narrow nation-state objectives have now come global concerns regarding future generations, such as environmental issues and, in particular, global warming. One of the consequences of this evolution is that diplomacy now requires many more skills, particularly in science and technology, than foreign affairs personnel are taught in diplomatic academies (Mayer, Carpes, and Knoblich 2014).

Diplomacy processes, practices and means: Diplomacy as discourse, communication, and negotiation between professional diplomats in dedicated settings is now only part of the diplomatic activity. Huge varieties of practices and means exist (ITU 2013), which are in constant development as cultural and intercultural mediations, as well as interpersonal relationships, playing an increasingly important part in diplomacy, including cooking and hospitality (Neumann 2011), fine arts, music (Ramel and Prévost-Thomas, 2018), and sports (Frank 2012), and are considered as part of the full range of diplomacy processes. Among the many examples of such diversification, public diplomacy and humanitarianism must be highlighted as forms of direct reach to people of foreign nations, most notably when usual diplomatic discussions avenues are difficult or entirely cut (e.g., public diplomacy has been practiced by the United States since the beginning of the Cold War; international humanitarianism intensively developed since the 1970s and 1980s, after the 'foreign humanitarian intervention' doctrine was developed by the 'French doctors' during the war in Biafra, and USAID was created in 1961). In recent years, there have been two particular moments where both public diplomacy and humanitarianism (sometimes in the form of development aid directly targeting civil society groups) played a major role before, during and after the event: the collapse of the Berlin Wall leading to the end of Cold War in 1989, and the Arab uprisings in 2011. In both cases, communication means were especially addressed and used, such as radio in the former case (Cummings 2009), the internet and, particularly, social networks in the latter (Clinton 2010). More generally, after the Cold War, the European Union exerted a significant role in stabilizing and democratizing Europe's Neighborhood through the use of its soft power, becoming in the space of a few years, a magnet of security with strong attractive power.

The mutations described so far are at the center of discussions[7] and analyses trying to clarify the concepts characterizing contemporary diplomacy, which remains fuzzy and overlapping. For instance, 'public diplomacy', 'science diplomacy,' and 'digital diplomacy' are often used interchangeably, without having been clearly defined and delimited (Cooper, Heine, and Thakur 2013b). While digital diplomacy for some aims at specifically addressing diplomacy objectives and practices in an age characterized by numerous digital innovations, public diplomacy and science diplomacy have an older history not necessarily linked to the digital era. Science diplomacy, in particular, has recently received renewed interest, with various attempts to define and flesh out the concept by, most notably, The American Association for the Advancement of Science (AAAS) and The Royal Society – UK National Academy of Science, and, as some chapters in this volume show, global internet governance may also be analyzed as one of a 'science in diplomacy' issues, i.e. global issues with scientific basis and the scientific/ technical aspects of formal diplomatic processes (The Royal Society 2010; Turekian et al. 2015). This undoubtedly demonstrates a political will to give a new impetus to diplomacy in the contemporary context.

As part of its innovative nomination of a 'digital ambassador to the internet giants' in 2017, Denmark has been particularly creative when coining a new concept of contemporary diplomacy to deal with Internet Diplomacy beyond its instrumental dimension of the use of digital means by diplomats: that of 'technological diplomacy', or 'Techplomacy', to use the portmanteau branded as a banner by the forerunner in the field, Casper Klynge, the former Danish ambassador in charge of these matters. We identify this promise of innovative diplomatic practices, one could even say of rupture, as a further development of an Internet Diplomacy research agenda.

More specifically, while, since 2017, other countries have nominated ambassadors in charge of digital affairs, almost none of them follow the Danish model. Under the French model, for instance, the digital ambassador is one of the twenty-one (as of June 2021) thematic ambassadors appointed by the French Ministry of Foreign Affairs. His mission covers the full digital governance spectrum of issues, including participation in bilateral, multilateral, and multistakeholder discussions on digital affairs. Given the current ambassador's public entrepreneur profile, who theorized the notion of 'Platform State' (Pezziardi and Verdier 2017), Henri Verdier's mission has extended to the development of Gov/Civic Tech tools tackling global issues, such as the fight against disinformation. In the Australian model, also adopted by Estonia and Finland, the role of the digital ambassador sticks more to classic regalian diplomacy, with cybersecurity issues being at the heart of the mission. Such different visions, models and strategies of 'techplomacy' vary considerably with the underpinning political orientation of the government defining them and appointing the ambassador. The diplomatic style and practices are also shaped by the ambassador in place, and the person's own background and culture. As a matter of fact, the new Danish digital ambassador, Anne Marie Engtoft Larsen, took office in October 2020. One would think that Danish 'techplomacy' would continue on the same line, but, in the meantime, the Liberals have been replaced by the Social Democrats in the affairs of the country, and the words of the new Danish Foreign Minister, Jeppe Kofod, in the nomination press release dated 22 August, 2020, suggested that a political shift may be coming, promising a 'Techplomacy 2.0'. It remains to be seen whether the 'Tech for Democracy 2.0' initiative launched by the Danish government in June 2021[8] will indeed lead to a new strategy and a new start for the Danish technological diplomacy.

While possessing various strengths depending on the 'Techplomacy' model, polylateralism and the role of private players as a power in international relations remain an inexorable trend, to the extent that a strong 'porosity' exists between the tech giants and the diplomatic world: for instance, Nick Clegg, the UK Deputy Prime Minister 2010–2015, joined Facebook in 2018 as Vice President for Global Affairs and Communication, and Casper Klynge, the first Danish Tech ambassador 2017–2020, joined Microsoft in 2020 as Vice President for European Government Affairs. Exploring all such cases and identifying whether this trend is simply 'revolving doors' as usual or a true 'Techplomacy' and 'Diplotech' encounter leading to deep mutations of diplomacy practices and outcomes must become an important strand of an Internet Diplomacy research agenda.

ORGANIZATION AND CONTENTS OF THE BOOK

As a contribution to this overall research agenda, the main research questions that this volume aims to answer are: can we see an emerging Internet Diplomacy as a new diplomatic field? If so, what do we mean by Internet Diplomacy? What are the diplomatic challenges around the governance of the internet? Does Internet Diplomacy develop new models and practices in the context of diplomacy? With this book, we thus approach Internet Diplomacy beyond the instrumental use of digital technologies for diplomatic practices.

As already discussed, the book doesn't address the use of digital means by diplomats to practice a kind of 'Public Diplomacy 2.0', which is explored by scholars under 'digital diplomacy' studies (Manor 2019; Bjola and Zaiotti 2020). This volume contributes to both the scholarly conversation and the global policy developments in the field by addressing how global internet governance, including cybersecurity policies, could be framed as an Internet Diplomacy area. As a matter of fact, even beyond its cybersecurity dimension, global internet governance in all its dimensions and areas could be addressed, analyzed and assessed as a 'science diplomacy terrain' and means of 'soft power' (Nye 2004), where scientific and technical expertise join forces with political influence and diplomatic action to address global challenges. This is particularly true considering the importance of technical experts and technical organizations, recognized as a stakeholder on its own in the multistakeholder regime of global internet governance.

With this book—which is, to a large extent, the unfolding of a conversation that started in 2017 among a network of scholars interested in exploring global internet governance actors, regulations, transactions and strategies and gathered for the first edition of the GIG-ARTS conference[9] to address global internet governance as a diplomacy issue—we have the ambition to unfold the concept of Internet Diplomacy by taking into consideration both the above-mentioned peculiarities of the emerging diplomatic practices in the governance of the internet and their outcomes in terms of normative transformations at the global, regional, and national levels. In particular, with the goal to understand and formalize Internet Diplomacy across all its dimensions and from multiple interdisciplinary perspectives, this book includes contributions addressing diplomacy around the international debate on the governance of the internet. A special emphasis is given to the role of the European Union and its member states in a field historically dominated by the US voice in the debate, due to its crucial role in the history of the internet, but also because of the leading position of the US internet giants in the global digital market. This book approaches the topic from an interdisciplinary perspective, by including contributions from leading scholars in the field of internet governance, approaching the topics from multiple backgrounds and disciplines, combining complementary novel theoretical approaches and empirically grounded research in the field of the governance of the internet as a diplomacy issue. This volume is, therefore, composed of ten chapters organized into three parts.

Part 1 explores how internet governance may constitute a (new) diplomacy issue in its own right, with the first three chapters respectively putting internet governance in the long-term perspective of the historical developments of diplomacy (chapter 2 by Yves Schemeil); analyzing it in relation with the two concepts of global governance and diplomacy, while taking into account specifics of the internet governance field, first and foremost the technology aspect (chapter 3 by Katharina E. Höne); and tracing how it has been politically constructed with different definitions, scopes and visions by the various stakeholders participating in the ten-year review process of the United Nations World Summit on the Information Society (chapter 4 by Mauro Santaniello and Nicola Palladino).

Part 2 more specifically analyzes whether and to what extent internet governance could serve as a science diplomacy instrument, exploring its opportunities and pitfalls through the relationships between science and authority, showing how the latter characterizes internet governance arrangements (chapter 5 by Robin Mansell); the comparative perspective between the US and Europe cases is explored in detail, with a focus on public diplomacy (chapter 6 by Nanette S. Levinson) and on cybersecurity policies (chapter 7 by Francesco Amoretti and Domenico Fracchiolla).

Part 3 presents four case studies that address in deeper detail, through empirical research, how internet governance diplomacy may be a means for the diffusion of values, norms, and policies from some regions of the world to others where internet governance and other digital regulation is less developed, and to what extent this may impact national sovereignty. Provided cases studies cover transatlantic free trade agreements and data flows (chapter 8 by Maria Francesca De Tullio and Giuseppe Micciarelli), the liberalization of telecommunication markets and its impact on transnational surveillance (chapter 9 by Claire Peters), privacy and the right to be forgotten in Latin America (chapter 10 by Jean-Marie Chenou), and international policy diffusion in the fields of copyright and privacy (chapter 11 by Krisztina Rozgonyi and Katharine Sarikakis).

With his chapter on 'Undiplomatic Ties: When Internet Blocks Intermediation', Schemeil (2022) opens the first part with a historical approach to question whether, how and to what extent the internet and its governance might—if it has not already done so—transform diplomacy, seen as relying on intermediation. Considering the highly privatized character of the internet ordering and the diversity of non-state actors intervening in its multistakeholder governance, the questioning focuses on the evolution of two main aspects of diplomacy: its practice as a formal communication process conducted by professional ambassadors in conventional settings and following established rules; and its organization as a multilateral or bilateral negotiation process between states. The chapter then examines two hypotheses: (a) the internet as a shortcut to classical diplomacy; and (b) that internet governance could only be effective through professional intermediation. To explore these two extreme situations, the chapter provides historical developments of diplomacy and up-to-date analysis of internet governance processes, both of which are highly relevant for the reader to understand the dialectics of two apparently mutually exclusive processes and antagonistic concepts.

Then, in chapter 3, Höne (2022) digs deeper to conceptualize the relation between diplomacy and governance in the internet field while avoiding one concept subsuming the other. In terms of methodology, the chapter suggests thus to refer, under the governance concept, to institutions and the set of rules and norms they define and apply, and to consider diplomacy when dealing with actors and their practices. The approach particularly fits the internet field, where new categories of actors have gained a seat at the table of negotiations, making multistakeholderism an especially prominent feature in the related discussions. These new blocks of actors include private ones, such as the business sector and civil society in the same way as in some other areas, as well as individuals who may bring their expertise to the discussions, but also artifactual ones, namely the internet infrastructure itself and its protocols, leading to a situation where technological developments transform both the nature, the substance and the outcomes of the debate. Here again, the dialectic relation between diplomacy and governance is highlighted, through several examples in the internet field.

With chapter 4, Santaniello and Palladino (2022) complete this first part by providing evidence that internet governance discussions are truly a diplomatic process that aims at making different definitions, scopes and visions from various stakeholders coexist, cooperate, and interoperate. The authors proceed through discourse analysis of different stakeholders' contributions to the WSIS+10 review process, ten years after the first World Summit on the Information Society held by the United Nations in 2005. The authors identify four coalitions ('neoliberal', 'sovereigntist', 'constitutional', and 'developmentalist'), providing the list of their members among the contributing stakeholders according to their own classification, and identifying the main contentious issues between them. These coalitions are, of course, ideal types defined for the sake of this analysis, and provide an empirical illustration of where main tensions and contentions lie in internet governance and how they are expressed by involved stakeholders. With this contribution, the authors shed light on the process by which actors coalesce around common narratives and eventually produce discursive orders.

Opening part 2 with her chapter on 'Science Diplomacy and Internet Governance: Opportunities and Pitfalls', Mansell (2022) starts by exploring how internet governance might be a field where science diplomacy can be deployed, especially considering its highly technological features. To this end, in particular, she examines the relationships between science (and scientists) and authority, both constituted and adaptive, where the former is predominant in science diplomacy and the latter characterizes internet governance arrangements. Here again, controversies and conflicts are traced and tackled, and attention is particularly paid to situations where academic researchers in internet governance engage in tackling socio-political challenges associated with the digital environment, and to the authoritative status of research evidence in situations where it may affect the interests of certain stakeholders. Further, taking into account the political economy of digital markets and the increasing powers of digital platform companies and their influence on the regulation and governance of the field, the author highlights how challenging the protection of citizens' interests becomes in such an environment. Science diplomacy is then discussed as a potential means of influencing diplomacy, in order to channel respect for democratic values and fundamental rights in internet governance discussions.

Still analyzing internet governance as a potential science diplomacy arena, Levinson (2022) focuses on the case of the United States with her chapter on 'Crafting Science Diplomacy In Comparative Perspective: The Case of US Internet Governance', in view of providing elements for a comparative perspective and paying particular attention to relevant cross-cultural communication and public diplomacy research and writings as well as to the public and science diplomacy practices of the United States. The author revisits, in the internet era, the US long-standing tradition of public diplomacy, arguing that the development of new media has transformed what she identifies as 'diplomacy places'. This resonates with the rise of new diplomacy actors in internet governance multistakeholder processes. Then, the chapter explores in which ways science diplomacy follows the same path, especially in the internet governance field. Comparative elements from Spanish public and science diplomacy are provided. The chapter concludes with some directions for the development of this research area.

Continuing the comparative approach, Amoretti and Fracchiolla (2022) address, in their chapter on 'Modes of Internet Governance as Science Diplomacy: What Might the EU Learn from the US Cybersecurity Policy?' more specifically cybersecurity as one of the most important issues in the internet governance field, and probably the one that most immediately illustrates how internet governance constitutes a (science) diplomacy issue. The chapter provides, with many examples, a thorough comparative analysis of US and EU cybersecurity policies, and examines common and diverging elements in their respective internet governance strategies and policies in this regard; it also analyzes the state of transatlantic cooperation in this field. In conclusion, the authors consider different scenarios on how the EU cybersecurity policy may develop in the future. Concluding the part on internet governance as a science diplomacy area, this chapter focusing on cybersecurity constitutes the perfect transition to part 3, which provides case studies on internet governance diplomacy.

The first case study proposed in part 3 deals with free trade agreements and their impact on internet governance. In chapter 8, De Tullio and Micciarelli (2022) analyze the role and position of the EU and other state and non-state actors involved in the negotiation of free trade agreements, that led to what they call 'free trade governance'. They also discuss how new public-private institutions created by the neoliberal design of free trade agreements may affect internet governance, in that they generate a shift of power on a transnational scale, since private subjects act as real negotiators, having an authoritative substance behind their formal corporate nature. The authors more specifically consider the Trade in Services Agreement (TiSA, whose negotiations were paused following the US presidential elections in November 2016) as a case study, to analyze how it affects crucially both political institutions and internet governance issues such as privacy and personal data protection, as well as network neutrality and transnational data flows. The study is framed in the broader theoretical background of constitutional law, governmentality studies, and a Foucauldian perspective.

In her chapter on 'National Sovereignty, Global Policy, and the Liberalization of Telecommunications Markets', Peters (2022) addresses, as a second case study, the consequences of the privatization of the telecom market on surveillance, with the evolution of the regime of lawful interception in Europe and of international cooperation of law enforcement authorities. By focusing on the issues of sovereignty, international cooperation, and public-private cooperation, the chapter analyzes it as an international relations issue, addressing inter alia the important cyber-normative field of international human rights. The author establishes a causality link between the privatization of the telecom sector and the evolution of surveillance legislation and shows how the globalization of private telecom operators led to problems with the enforcement of national laws that could only be solved through diplomacy and international standardization. An enlightening example of 'law enforcement authority diplomacy' is provided with the reported FBI initiative. The detailed analysis provided in this chapter constitutes an essential tool to understand the never-ending developments in national and transnational surveillance based on telecommunication data, and the increasing issues of national sovereignty in the field.

'In This Bright Future You Can't Forget your Past: Debating the "Right to Be Forgotten" in Latin America', says Chenou (2022), who presents in his chapter an example of law and policy diffusion, that of the EU 'Right To Be Forgotten' (RTBF) in Latin America. He analyses how the issue was debated and implemented in four large countries (Argentina, Brazil, Colombia and Mexico), with regard to two possible conceptions of the RTBF: that of the EU and that of the US. Tracing back to the 'Costeja' case that initiated the European Court of Justice's decision on the RTBF and the subsequent EU legislation, the chapter presents and discusses the US and EU's different approaches to internet intermediaries liability, and highlights the difference in the ways of approaching the role of search engines, in this case, that is at the heart of the two competing visions of the case (data controller in the ECJ decision vs. internet intermediary in the US vision contesting this decision) and, consequently, the different vision of the applicable law (EU data protection legislation vs. US—as the search engine country—intermediary liability regulation). The chapter then discusses in detail the elements and tensions in this debate in the four considered Latin American countries, as regard the perspective to adopt in their legislation. Showing how these discussions were held in the shadow of EU and US (public) diplomacy and influenced by them, this chapter provides an important example of policy diffusion process and the role played by academic researchers and civil society actors in it.

In the closing chapter on 'Policy Diffusion and Internet Governance: Reflections on Copyright and Privacy', Rozgonyi and Sarikakis (2022) extend the elaboration of international policy diffusion processes in the field of internet governance, analyzing more specifically two case studies related to privacy (with the notion of informed consent) and copyright (with the concept of graduated response). The chapter focuses on policy transfer as a matter of foreign policy and investigates structural factors that influence this process, in particular the role of citizens' activism in international policy making. With these two selected case studies, the authors consider two emblematic 'sites of struggle' among the many characterizing internet governance contentions, that affect citizens' everyday life and their rights to access, use and generate internet content, impacting their very fundamental freedoms. As the authors argue, these cases entail policy principles that reflect ideological dispositions about the role of the state and the market, the role of the citizen as an actor and, ultimately, even the state of fundamental rights in a volatile world.

NOTES

1. As it is impossible to cite here all the relevant work in the internet governance field, this reference is provided as an anthology of such work by a number of authors covering almost all aspects of the global internet governance studies.

2. See UN WSIS related information on ITU website available at: https://www.itu.int/net/wsis/.

3. See UN GGE webpage at: https://www.un.org/disarmament/group-of-governmental-experts/.

4. See UN OEWG webpage at: https://www.un.org/disarmament/open-ended-working-group/.

5. Such as, for instance: illegal and harmful content vs. freedom of expression; state surveillance, private companies abuses, and various forms of cybercrime activities vs. the protection of privacy and personal data; intellectual property rights vs. a vision of internet content as commons to foster education, knowledge, innovation and global development.

6. See the IGF website at: https://www.intgovforum.org/.

7. See for instance discussions organized by The AAAS Center for Science Diplomacy (https://www.aaas.org/programs/center-science-diplomacy), The USC

Annenberg's Center on Public Diplomacy (https://uscpublicdiplomacy.org/) or The Diplo Foundation (https://www.diplomacy.edu/), or at the Clingendael Institute (https://www.clingendael.org/), to only name a few.

   8. See https://um.dk/en/foreign-policy/tech-for-democracy-2021/.

   9. See presentation of this conference series at events.gig-arts.eu.

I

INTERNET GOVERNANCE AS A DIPLOMACY ISSUE

Chapter Two

Undiplomatic Ties

When Internet Blocks Intermediation

Yves Schemeil

Diplomacy relies on intermediation: it channels indirect communication between governments. The Internet is a medium that directly connects nongovernmental actors. Historically, bilateral meetings paved the way to multilateral forums on which diplomats addressed international issues as public representatives of their country. Now that Internet governance has recently been privatized the future of state-led multilateralism and diplomacy is uncertain. Diplomacy is shortcut by coarse undiplomatic behavior and multilateralism is jeopardized by self-sustained decentralization. What's next? Will negotiations run by ambassadors working in formal venues remain center stage rather than definitely losing relevance? Will the newborn Internet with its egalitarian design complement and even relay classical diplomacy instead of killing it?

The end of interstate diplomacy cannot be inferred from the proliferation of bypasses. Diplomats are curious and adaptable: 'epistemic motivation' makes them open to new information, wherever it comes (Rathburn 2014, 6).[1] On the other hand, 'non-state diplomats are now more widely recognized as populating a scene that might in fact have far more "techno diplomacy" than ever before' (Acuto 2017, 93). Moreover, isn't communication a shared goal of agents in the old and new age of diplomacy (Trager 2017)?[2]

THE RESILIENCE OF DIPLOMACY

As an alternative to war, negotiation is as old as the first documented societies. Five thousand years ago, a delegation of elders from a besieged city met the chief commander of the enemy's army to pledge loyalty or pretend they could fight to death. As in Thucydides's famous Melian dialogue, so cherished by realist scholars, the options were 'surrender or perish'. Although some talent was needed to tame fierce conquerors, in ancient times diplomatic skills were either unknown or useless.

When diplomacy emerged from the chaos of the European or Asian battle-fields in the eighteenth century, the range of options was enhanced, as was the set of capabilities needed to reach an agreement. The new philosophy was: 'negotiate and concede'. Bargaining, trade-offs, and compromises were now on every negotiator's horizon. Secrecy and moderation replaced publicity and arm-twisting. Far from representing their constituency as the wisest men of a community of people from whom they were distinct by age only (as in the ancient Orient), European diplomats were selected by governments among the ablest professionals of their time. They were specialized, and they praised confidentiality. Their capacity to convince others relied on three aptitudes: sound knowledge of precedents, correct assessment of the actual balance of power, and ability to distinguish between bluff and determination. However, persuasion also depended on fluency in the use of exaggeration, double-talk, and lies—only thinkable behind close doors, hidden from the eye of any beholder. And this is exactly why, in multilateral organizations, governments refuse access to NGOs, the press, and even nonaccredited academics (Schemeil 2004).

Confidentiality at Risk

The invention of the Internet puts such certitudes upside down: the web nearly ruins confidentiality. Diplomatic events are commented by count-less bloggers; finger-pointing is the only sport in town. Additionally, this participatory democracy makes time and space shrink, leaving no room for regret or remorse, backstage negotiations, and time for assessing the impacts of successive decisions before accepting to pick the less harming. Because negotiators are evaluated online, they aim to perform well. They try to avoid prosecution by an international court or bashing by the tribunal of opinion. Hence, three of the major ingredients of a successful negotiation (*immunity*, *secrecy*, and relative *independence* from constituents) are no longer granted.

What is in jeopardy is the classical system of face-to-face/eyeball-to-eyeball diplomacy in a 'promiscuous and partially closed . . . microcosm' where 'groupness and nearness' are the rule, and which was 'the lynchpin of international politics' for centuries (Holmes 2013, 829). Even between two states, negotiations often involve more than two governmental teams. They address 'multiple audiences . . . and different publics simultaneously' (Pouliot 2011, 546, 555). When the remotest and less concerned contributors can participate to an extremely inclusive decision-making the legitimacy of multilateral negotiation is weakened. Other things being equal, the model may therefore become a bygone in less than a decade.

Does this grim perspective change when diplomacy is viewed as practice and not as an institution? At first glance, the 'practice turn' in international studies gives a different account of the future of diplomacy, mainly because its classical version was actually more 'modern' than assumed for decades by mainstream science. According to conventional wisdom, diplomacy is made of rules and institutions like treaties and protocols of negotiations. By contrast, diplomatic practice is permanently reinventing the framework of diplomatic interactions, increasingly made of 'mundane socio-processes' (Pouliot 2016) instead of official codes of conduct determining behavior 'by the book'. Practitioners who 'are never placed twice in the same situation' are busy with 'the achievement of infinitely diversified tasks' that requires 'improvisation' and 'virtuosity' (Cornut 2018)—two skills that are displayed offline and online altogether. Therefore, diplomacy becomes less hierarchical, less formal, and less routinized (Cornut 2018, 724). Such a trend would have occurred with or without the Internet[3].

From Theory to Practice and Return

Similarities between diplomats and geeks explain why antagonist schools of thought do not entirely disagree on the resilience of diplomacy.

Whether it is a slow-changing institution rooted in history or a rapidly evolving practice boosted by technological innovation diplomacy remains framed by a number of constraints. The debate can therefore be reformatted as follows: Will transparency, instantaneity, rating, and boldness kill 'gentlemanly diplomacy inherited from a state-centric world'? Will orthodox diplomacy overcome the challenge raised by 'various heterodox forms of political intercourse'? Will instant posts from anonymous geeks replace face-to-face meetings in which negotiators can decipher each other's feelings and read the clouded intentions of others, at the risk of compromising the capacity to clarify positions and solve intractable issues (Wong 2016; Holmes 2013[4])?

As stated by most observers, 'the rise of nonstate actors on the diplomatic stage' induces a loss of governance in world affairs—although not necessarily less capability to negotiate, because 'unofficial diplomacy' also contributes to the atonement of adversaries (Davidson and Montville 1981). Despite this, the old system resists: classical diplomats are not as traditional as alleged; they have progressively departed from the patterns established centuries ago in non-digitalized times. Hence, it is likely that 'old and new practices coexist in a mutually constitutive relationship' (Sending, Pouliot, and Neuman 2011, 528–29).

Outside the 'diplomatic stage', however (a situation little explored by mainstream internationalists as well as promoters of a theory of practice), coexistence between old and new actors, as well as between established behaviors and unprecedented codes of conduct, is less and less likely. The emergence of a 'multilateral diplomat profile' (Pouliot 2011) may be compromised if permanent representatives could be rivaled or replaced by non-permanent and non-representative agents.

In recent years trials to sideline embassies did not go unnoticed. As a last blow against multilateral diplomacy, the World Trade Organization was created in late 1994—one step before digitalization went global. Coming after decades of tough negotiations within the GATT, the 1995 birth of the WTO expressed indifference if not hostility towards the two intergovernmental organizations in charge of global trade (UNCTAD and WIPO). When ICANN emerged on a Californian campus, it ignored the two IOs regulating telecommunication and intellectual property (ITU, confined to the provision of a technical standard setting) and WIPO (despite increasing concerns for the protection of copyrights and patents in a world of 'free software'). However, recent evolution shows that it is not so easy to get rid of the classical model of diplomacy. Although the WTO was advocated as a new kind of organization in which judges would prevail, it nonetheless remains a diplomatic forum.

Therefore, diplomacy is consubstantial to the world we know because politics is the art to live with people we do not like. It is therefore likely to remain central in interactions with foreigners in the next decades. Secondly, classical diplomacy may be partially obsolete but Internet generations themselves are indebted to its achievements: it has now become 'spectral' (Der Derian 1999). Thirdly, attempts to shortcut embassies and to rely on contracts rather than institutions require more organization, not less: decentralizing participation requires a sophisticated centralization of voting procedures.

In the rest of this chapter, I test two conjectures. According to the first, the Internet shortcuts classical diplomacy; on the contrary, the second predicts that Internet governance will never be effective without professional intermediation.

DOES THE INTERNET SHORTCUT CLASSICAL DIPLOMACY?

From the earliest times special representatives acting as delegates of a ruling body negotiated cross-borders arrangements. The Sumerian League (third millennium) and the Peloponnesian League (fifth century) are among the best documented. The Chinese-led peninsular tribute system that was established at the turn of the first millennium has had less coverage; we nonetheless know with enough evidence that it favored the emergence of diplomacy in Korea and Japan. Meanwhile, Persia, Egypt, and the Ottoman Empire carved countless agreements in the Middle East. Centuries later, the sixteenth, seventeenth, and eighteenth centuries were full of peace conferences between great powers of the time. These steps prepared the birth of the 1815 Congress of Vienna and its many innovative achievements, including interstate organizations. Therefore, diplomacy and multilateralism as we understand them were born in 1648 and 1815, respectively, and they triumphed in the twentieth century.

Classical Diplomacy

What is the 'essence' of diplomacy? Conventionally, its role is to channel communication, hence enabling a dialogue between partners who would otherwise be 'estranged' from each other. Such a view conflates diplomacy and mediation: diplomats are at the interface of peoples whose relationships are both *alien* and *asymmetric*: other cultures always look strange; philosophical conceptions differ. This world is separated or not from the other world (Der Derian 1987). Political systems give sovereignty to the people (in the US), to a parliament (in the UK), or to a party (in autocratic regimes).

Of course, diplomats learn to get along together. In one of their latest transmutations 'as stabilizers of problematic and problematized identities' they produce and reproduce their professional identity (Sharp 1999, 48–49). This is the only way to solve a contradiction at the roots of their vocation: asserting that diplomats who are different in culture, language, and political principles can nonetheless have a dialogue thanks to the same negotiating talents. A solid diplomatic culture has been 'arising out of the experience of conducting relations between peoples who regard themselves as distinctive and separate from one another' (Sharp 2004). The keyword here is 'conducting': geeks do not feel conducted at all; allegedly, online relations can proceed without any benevolent mediator.

Due to cultural and ideological variety diplomats must remain *neutral* and *equals* to reach an agreement; they must refrain from expressing value judgments, and never make any use of force; they have to express concern for *face-saving*. A rich 'repertoire' of 'subtle' tools 'serves as a micro-foundation for conflict avoidance among diplomats and statesmen'. This facilitates de-escalation, and makes negotiators feel equal (Nair 2019, 674; 677). Face-saving is a clear comparative advantage of professional diplomacy over unframed amateurs' discussion, which is full of asymmetric losses of face occasioned by blunt 'naming and shaming' on social networks. While real negotiation sites allow negotiators to adopt red lines, ban sources of embarrassment, move them to backstage, or defer them to junior diplomats (Nair 2019, 688), virtual sites cannot.

With so many buffers against a loss of temper, diplomacy depoliticizes hot issues. It is not only the antinomy of war: it is also the enemy of ideology. This means that emotional attachments to one's beliefs are concealed, because diplomacy is not mere statecraft, but a vocation and a career (Sharp 1999). Diplomats know how good they are at using rational arguments in order to reach a comparatively advantageous position during a negotiation. But they cannot be too bold or too aggressive: even when they are bargaining, they must avoid open threats and coarse promises. They are not philosophical subjects committed to values; they are mediums, i.e., instruments conveying the will of others.

New Diplomacy

On the contrary the Internet is the locus of passions and vocal commitments. Although its end users are supposedly equals, the reality is: they are not. A new hierarchy of 'followers' and 'likers' gives more leverage to 'influencers' and less to 'friends' (as in 'friends of the chair'). None is neutral, indeed: hot debates compel to bluntly and boldly take sides.

Furthermore, relaxing some traditional conditions to make communication unequivocal may be lost: initially, diplomacy was associated to four qualifications ('*authenticity*', '*accreditation*', '*esoteric activity*', and '*linguistic precision*'). The first is in every geek's DNA. The three other properties are little compatible with the philosophy and practice of the web.

Take accreditation: being accepted as a legitimate discussant on a virtual forum is unconditional. True, a maverick might be expelled from a forum when transgressing the code of conduct. In such a case, the site administrator simply raises an alert flag—a sort of accreditation by default for all the others who respect the rules. However, it is not a positive recognition of the capacity to debate.

As for publicity, while diplomacy is confidential if not esoteric, Internet discussions are clearly exoteric. This remains true despite identification cues, firewalls, and passwords, regardless of the respective frequency of leakages (in diplomacy) compared to hacking (on the web).

Finally, language clarity is undoubtedly lacking online: e-users are more than tolerant towards weak syntax, lack of style, confusing vocabulary, and incorrect use of words—not to speak of automatic proofreading, which may result in fancy if not counter-productive messages[5].

There are other sources of change in diplomatic ways. Technical innovation opens new opportunities to express one's views faster than with the good old telegrams. Rulers now communicate directly and instantaneously with each other and with their respective peoples. In 2016 and early 2017, for instance, the US State Department was technically prevented from working normally by the new president's practice (debating with his close entourage, if not family; tweeting blunt statements based on rumors launched by ideologically biased sources; and, possibly, making public statements on top of the head). Trump's language was undiplomatic in many ways, blunt, coarse, or even insulting, breaking with a tradition of caution and precautions that can be traced to 1648 Westphalia. He and other 'strongmen' have made a strategic use of mass disinformation and crowdsourcing—whereas real diplomats rely on secrecy and parsimony. Reinforcement of false evidence, pseudo facts, and incorrect arguments raised by people unaware that their information comes from the same root add to the current demise of expertise. If leakages are not accident but calculated tactics, then we loose on both realms: lack of secrecy makes room for premature statements, which are sadly not easy to suppress; excess of publicity trivializes what should remain a historical moment. At any rate, a picture of a politician tweeting on a sofa will never be an acceptable substitute to historical photographs and paintings depicting the end of international peace conferences in famous venues (be it Vienna, Versailles, or Camp David).

A Mix of Old Hats and New Geeks' Diplomacy?

Diplomacy evolves fast. Traditions vanish; ceremonials are increasingly confined to exceptional moments like commemorations or national funerals. Confidentiality is no longer kept, as if diplomats were psychiatrists collecting the confessions of their patients. Contrariwise, they favor leaks and press release event when this may be embarrassing or harming for their partners.

One symbol of this loss of diplomatic ethos is the opening up of closed sessions. So, we may wonder what is lost when 'greenrooms' (i.e., places where staunch conflicts are cleared behind closed doors) become glass houses?

Firstly, the room of maneuver offered by the distinction between 'front' and 'backstage'; 'formal' and 'informal' meetings; 'press releases' and 'leakages' from unknown or unnamed sources simply disappears. When everything is put on the table, successive stages of meticulous drafting that had laboriously led to a hard-to-achieve agreement are concatenated into a consensual package, visible to all with no regard for confidentiality and anonymity. For example, when WIPO started publishing all intermediary texts short of a final agreement, in which brackets still to be waived at each stage of the negotiation process were left, it also limited the ability of permanent representatives to use a double standard according to the audience addressed—fellow diplomats during the negotiation process or compatriots back home.

Secondly, the legitimacy of diplomatic outcomes is impeded when countless self-claimed stakeholders can attack it any time anywhere on cyber forums. As a consequence, immunity that was conditioned to self-restraint in public is gone. Within the WTO, representatives of states opposed to a consensual agreement in the making may frankly tell their peers that they personally side with them rather than supporting their own government and not be sanctioned for such behavior because their personal position will never be known outside the room. Once debates are online, any deviance becomes public and can be criticized; those who dared to express a dissenting opinion are lambasted if not repatriated and possibly fired.

Thirdly, a nascent 'digital culture' makes progress in every IO. In the first two decades of the WTO, permanent representatives used an arsenal of trade-offs and tit-for-tat tactics to address technical issues. Instead of solemnly drafting covenants as in the GATT time negotiators now attend conferences with a cool attitude in casual clothes; they reach agreements in a short span of time with peers from other countries, independently of other stakes negotiated in parallel elsewhere. They build 'coalitions of the willing' with those ready to get along together without waiting for latecomers who must accept to catch up with any progress made in the meantime. The emergence of this new generation contributed to the dissolution of former coalitions like the Quad (the club of the wealthy); or the alliance of the developing countries (the G 22 or 77 groups of the less affluent). It has accelerated the obsolescence of diplomatic norms shared from the earliest years of the organization—like the 'single undertaking' norm, i.e. 'nothing is agreed until everything is agreed' (meaning that the outcome of the negotiation process is a package that you can either take or leave but that you cannot split into separate parts). This may explain the collapse of the 'round' system—as evidenced by the never-ending 2000 Doha round.

Since the WTO is now addressing e-commerce issues, this new generation is particularly active (and successful) in the realm of digital trade. Here, the overlap between new norms of behavior and a new field of negotiation brings in quite unexpected groupings, like the 'Friends of the e-commerce for development' (Argentina, Costa Rica, Sri Lanka, RK, Kenya, and Nigeria) or, weirdly enough the MIKTA (Mexico, Indonesia, Republic of Korea, Turkey, Australia). The idiom used by such a geek generation within exclusive circles of mutual recognition and mutual interest in e-commerce is Internet-friendly, color-blind, and cosmopolitan. Of course, negotiators want to go fast. They leave little room for second thoughts, neglect possible overlaps with other stages or arenas of discussion, and cannot wait for policy turns that take decades to reach (since adversaries must be patiently rallied to a cause, if not bought out with pledges to help them on another stage). This is also the end of the iterative process of negotiation at the roots of classical diplomacy[6]. However, another continuum is drawn between amateur and professional diplomacy.

On the Internet, the slight boundary between formal and informal, *official* and *off* tends to vanish. A core achievement of diplomacy at the outset— the clear-cut distinction between appointed delegates and nonaccredited people—is now blurred. This inevitably occurs when official rounds of negotiation (or 'track 1', from which we have extrapolated the characteristics of classical diplomacy) are losing ground to 'track 2' channels (Montville 1991a). Motivated volunteers, famous artists or respected intellectuals side-line 'Old hats' (i.e. heads of delegation wearing black coats, jackets, hats, and ties)—regardless of existing dress codes, speech turns, and ways to talk[7]. Nonetheless, the growing influence of nonprofessional diplomats has not stopped the increasing demand for representatives mandated by their governments (Sharp 1999: 42–44). Therefore, whenever professional diplomats are stuck into quagmire, amateurs can relay specialists; but they cannot replace them in ordinary circumstances. Being online or off the record will coexist to some extent for a long time.

Perhaps diplomacy could have changed with or without the Internet. True, states representatives to foreign countries or multilateral negotiation are increasingly flanked by occasional diplomats—chefs, musicians, movie stars, and academics. Great powers involve scientists in their foreign relations, as aides or substitutes to ambassadors, practicing what they call 'science diplomacy'—neither 'science for diplomacy' (science as the pursuit of diplomacy by other means) nor 'diplomacy for science' (diplomacy as a way to promote national scientific achievements). Among such 'informal diplomats,' scholars are said to be the most independent compared to sport champions. They nonetheless collaborate with official ambassadors, whose outreach in foreign countries is proportionally increased, opening up their own networks of influence to the benefit of their home state (Fähnrich 2017).

Inviting scientists to call at embassies increases truth dissemination because the Internet makes open source possible, but it may impede diplomatic confidentiality. Suffering themselves from bureaucratic routines and top-down policy decisions scientists as occasional diplomats are in line with promoters of a free web. As geeks do in a cyber world where cosmopolitans outnumber chauvinists, internationalized scholars believe in mutual understanding across boundaries. While official delegates defend the national interest of their home state, their fellow scientists want economic returns from their assistance in governmental operations. Rather than looking for a direct political influence

Table 2.1.   Changes in diplomatic structures and agency

	Classical diplomacy (17th onward)	Diplomacy at the age of the Internet (since the nineties)
Legitimacy	Immunity	Challenged
Legality	Law/Treaty/Constitution	Bylaws/Contract/Terms of Service
Activity	Confidential	Transparent
Decision-making	Slow/Hierarchical/Formal	Fast/Egalitarian/Informal
Reasoning	Considerate	Affective
Institutionalization	Organized (IGOs)	Unorganized (NGOs)
Governance	Mono-stakeholder/like units	Multistakeholders/unlike units
Sovereignty	Delegation	Stewardship

on foreign colleagues, scholars as diplomats opt for a strategy of minimal 'door-opening' to domestic policy-makers. They are satisfied when their findings are 'branded' abroad. Their norms and expectations are therefore consistent with Internet rules, albeit less consistent with diplomatic mores (Fähnrich 2017, 696; 699)[8].

Science diplomacy also illustrates a movement towards 'hyphenated diplomacy'—a sign that intergovernmental action is no longer the best way to accommodate aliens (Hayden 2011). An Internet-friendly *nongovernmental* diplomacy in the offing will draw on one of the major functions of classical diplomacy (i.e., representing the masses and not only the elites; defending their national and not only their professional identity). But it will also be 'more inclusive and less ubiquitous' (Hayden 2011). In the end a sort of 'integrated diplomacy' combining several sorts of intermediation (old and new, vertical and horizontal, general and sectorial, governmental and nongovernmental) will be born (Hocking et al. 2012).

So future diplomacy will be a mix of old and new types, somewhere in the middle of the continuum between diplomacy at the time of the telegraph and diplomacy at the age of the Internet that is summarized in Table 2.1.

CAN INTERNET GOVERNANCE BE EFFECTIVE WITHOUT ANY INTERMEDIATION?

Internet governance is an increasingly debated issue, due to the absence of regulation of online deliberating processes, to such an extent that '(s)ome have labeled the IGF as a *decision-shaping rather than decision-making forum*, or a forum where issues of Internet related policy are being discursively constructed' (Epstein, Roth, and Baumer 2014: 145, emphasis mine; IGF stands for Internet Governance Forum).

Here, 'governance' is not only *short of government* (Schemeil 2012), it is also *soft governance* compared to hardest cases like climate change. It may also look like *governance against governments*, since the latter are relinquishing power to private agents who attend IGF meetings[9].

Internet governance is less framed than structured partnerships by the confrontation of national interests. It henceforth relies on 'loose couplings' (Goffman 1983[10]; Pouliot 2016), i.e. partnerships that can be defined as examples of '*(u)nstructured* diplomacy in which government has *low* presence: roles and relationships are *fluid* and responsibilities *not clearly defined*. Rules and norms (are) *underdeveloped or absent*' (Hocking et al. 2012, 19–20, my emphasis). Tight or loose governance: this may be the major division between what Internet governance would have become if it had been assigned at the outset to an IGO (like ITU) handled by diplomats.

Loose Couplings

One empirical example of an area in which the concept of 'loose coupling' applies best is: 'policing the Internet' (Hocking et al. 2012). Loose couplings may surge from any sort of provisional networking since non-institutionalized nodes of relationships can either shrink or disappear as quickly as they have emerged. This novelty has a deep impact on our classical conception of diplomacy, when it is compared with another notion, 'pseudo diplomacy':

> Diplomats will need to re-evaluate the role models on which their activities are based. On one hand, this will involve *relinquishing claims to hold a privileged position in a hierarchical environment marked by special qualities such as secrecy*. On the other hand, it involves developing strategies for reconciling the needs for policy specialisms with the more traditional diplomatic roles—such as the ability to interpret cultures and mediate between them—that globalization renders more significant. (Hocking et al. 2012, 22, my emphasis)

Managing relations mediated by the Internet as well as the Internet itself as a pseudo-diplomatic tool: here are two examples of what the author calls 'policy *specialism*'. Such specialization seems nonetheless compatible with the *generalist* stance traditionally assigned to diplomats; hence, division of labor might work as well in the realm of intergovernmental negotiations and not only in the domain of Internet debates.

Perhaps though, changing attitude is a matter of adaptation to a new instrument. Internet technology may be diplomacy-friendly—as were once the telegraph and the telephone, which 'have been as much a part of the shaping of the diplomatic milieu' (Hocking et al. 2012, 48). Alternatively, it may entail dramatic changes in diplomatic practice.

Are such changes impacting intermediation processes more than did the incremental adjustments observed when new vectors of communication emerged in the past? If we now consider diplomats' contribution to Internet governance, and not diplomacy as a profession whose use is inevitably transformed by such a new and powerful tool, are our views about the relationships between Internet and diplomacy outdated?

Comparing the Internet to former vectors of communication shows that the social networks it made possible are radically new: they deeply transform the status of the people connected. Notwithstanding the number of applications enabling anyone to interfere in any discussion anywhere at any time, the difference between past periods and today does not stem from speed or scope (telegrams already traveled fast and far) but from the status of communicators. Yesterday, channels of communication were confidential, they conveyed classified material from ambassadors to ministers and vice versa; now, anybody can obtain the proceedings of a conversation, if not crash a Zoom meeting and join panelists online. This is exactly what Ervin Goffman once called 'third parties' trying to become authorized speakers: they nervously wait until the protagonists of a debate remark their impatience and turn to them for external approval.

However, we should not overestimate the difference between old and new diplomacy on one particular point, privacy, which has increasingly obsessed participants to Internet Governance Forums (Mueller 2016). Excessive departure from the classical patterns of confidential diplomacy has triggered a counterbalancing move towards more protection of web users' intimacy (Epstein, Roth, and Baumer, 2014).

As for the second question, building Internet governance was from the very beginning an experiment in transnational ruling. NGOs, local or international, have been privileged over IGOs, the US Federal administration has overwhelmed the EU, and California has prevailed over China as a major stakeholder of ICANN. All in all, an unstable network web was preferred to the hierarchical structure that has so far kept global governance stable under intergovernmental control.

One step below a full-fledged organization like ITU, ICANN is loosely formalized. It relies on guidelines, not binding rules as ITU does; it mixes voting board members and 'non-voting liaisons', members of the board who can be easier to remove[11]. It is at the center of a privatized and flexible network of governance, whose 'stewardship' and 'bylaws' compete with constitutions and international agreements. Finally, it opposes incomplete contracts to detailed and precise constitutional norms:

These core values are deliberately expressed in very general terms, so that they may provide useful and relevant guidance in the broadest possible range of circumstances. Because they are not narrowly prescriptive, the specific way in which they apply, individually and collectively, to each new situation will necessarily depend on many factors that cannot be fully anticipated or enumerated; and because they are statements of principle rather than practice, situations will inevitably arise in which perfect fidelity to all eleven core values simultaneously is not possible. The Board will exercise its judgment to determine which core values are most relevant and how they apply to the specific circumstances of the case at hand, and to determine, if necessary, an appropriate and defensible balance among competing values.[12] (ICANN website, Governance)

Moreover, ICANN cajoles its community of experts rather than spending time healing governmental wounds. It 'respect(s) creativity' and pledges to 'remain accountable to the Internet community'. Therefore, attention is given to the 'ombudsman', an advocate of rank-and-file or 'at large' people, who is dedicated to the primacy of fairness over rectitude and must 'ensure that those entities most affected [by any ICANN decision] can assist in the development process' of a policy measure. He or she also relies on '*shuttle diplomacy*' to mediate between the various layers of the Internet community[13], as evidenced by its organizational chart[14]. A glance at this chart shows how much private and open-ended the founders of ICANN ambitioned to be compared to public institutions like ITU (which ICANN calls, not without prejudice, an 'international treaty organization'). Actually, most of its bodies have a consultative rather than a decision-making power. With its 'task forces', 'advisory committees', 'nominating committees', and 'CEO', ICANN is a hybrid.

ICANN officers believe that expertise comes from everywhere and become 'crowd-sourced by the community' or stems from 'open community discussions'. Even within the Government Advisory Committee (the GAC), and long before recent pandemics, diplomatic conferences have been mostly replaced by 'webinars'. Departing from accepted vocabulary is justified by the specificity of the domain. If ICANN has a depoliticized technical mandate, it is to insure 'interoperability' and 'universal resolvability', and to guarantee that the Internet works in the same way wherever its end users are[15].

ICANN also reminds stakeholders that one of its 'core values' is to work with other units—a lip service to cooperation, which is also paid on every website. The difference with other international organization comes from the promise to '*delegate coordination functions* to or recognizing the policy role of other responsible entities that reflects the interests of affected parties'—a wink at the arbitration center of WIPO[16]. Such promise to share power with peer organizations is very unusual in the multilateral world.

Finally, some commitments made strongly point out by default the caveats of classical diplomacy and multilateralism (like: 'promot(ing) well-informed decisions based on expert advice'; 'mak(ing) decision by applying documented policies neutrally and objectively'; 'acting with a speed that is responsive to the needs of the Internet'; or respecting 'time limits' to prevent lifelong service). By contrast, board members are considered as corporate officers, belonging to an entity 'rooted in the private sector', which 'promotes a competitive environment . . . depending on market mechanisms'. As do staff people in a firm, board members have a right to 'continuing education'; but they are also warned against making 'repeated offenses' to their fellows or concealing conflicts of interests during and after their mandate; full attendance to scheduled meetings is encouraged; performance is understood as 'effectiveness' and not only as efficiency.[17]

Is ICANN a Meta-Organization or a Hyper Social Network?

Is ICANN an 'international organization' consistent with classical diplomacy, or is it a brand new type of organizational networks belonging to an emerging type of regulatory body?

After due consideration, ICANN could be the core node of a '*meta organization*' made of heterogeneous unlike units (like states, firms, NGOs, universities and think tanks) that are linked together through it. Its identity and autonomy would then compete with the identity and autonomy of each of its component, and this would, in turn, press to limit the number of participants to the most similar ones (Ahrne and Brunsson 2008, 2012). In such a case, classical diplomacy is needed to coordinate single organizations within the meta-organizational structure. By contrast, the inclusiveness at the roots of multilateralism, which is favored by a common diplomatic language, is endangered. However, meta-organizations are weaker than single organizations, a fact that favor 'recommendations', 'white books', and soft law (Ahrne and Brunsson 2012; Ahrne, Brunsson and Kerwer 2016)[18].

Alternatively, ICANN may be but a macro replica of the myriads of social networks that owe to the web their very existence and usually can do without any diplomacy—a sort of hyper social network. Diplomats are not necessary to mediate between grapes of interrelations, since mutual adjustment between 'friends' and the dichotomy between 'likes' and 'dislike' is mechanical—there is little to do about it. Besides, boundaries between organizations that are enmeshed within the web of Internet governance are not so sharp that they would require intermediation; they can be easily and instantaneously trespassed. In such a context, monitoring the implementation of decisions is difficult. The only way out of this trap to control the governance of the whole system is to reinforce similarity between the components and understate conflicts, against imperatives to open up to difference and smooth contradictions (Ahrne and Brunsson 2012; Ahrne, Brunsson and Kerwer 2016).

Whatever it is (a meta-organization or a hyper social network), does ICANN succeed better than the single-issue organization dedicated to tele-communications that is also involved in Internet governance, ITU? Central-ized power in a well-established organization socializes new members or new representatives to a shared diplomatic culture made of norms of behavior that have currency within the institution. By contrast, diffused competence within a 'minimalist organization' makes difficult to manage reputation and produce an organizational identity able to mobilize its members. To socialize them, any IO should be a 'cultural producer' of both identity and beliefs (Spillman 2017), able to convince others of its legitimacy and specificity. This, in turn, would require communicative skills. Since ICANN aim to facilitate com-munication, it may be taken for granted that its heads are better communica-tors and produce more original culture than ITU ever did (Jönsson and Hall 2003). However, lack of centralization and structural stability, due to the US Government and the IGF dedication to multistakeholderism[19] weigh much on its efficiency. This may explain why ICANN founding fathers and sponsors altogether opted for a non-structured and fluctuant epistemic community doing some sorts of peer reviews rather than a full-fledged intergovernmental organization.[20]

Eventually, rather than being the product of a meta-organization or a social network, Internet governance relies on a loosely knit network of heteroge-neous units put on an equal footing despite striking asymmetries, which is regulated by mutual adjustment processes. Among the major stakeholders of Internet governance, IANA (Internet Assigned Numbers Authority) was 'one of the Internet's oldest institutions . . . , dating back to the 1970s'.[21] Its rationale for endorsing such a complex system of management is stated as follows:

> Whilst the Internet is renowned for being a worldwide network free from central coordination, there is a technical need for some key parts of the Internet to be globally coordinated, and this coordination role is undertaken by us (i.e., IANA/PTI).

On the same website, ICANN is described as 'an internationally-organized non-profit organization set up by the Internet community to coordinate (various technical) areas of responsibilities'. ICANN implements the IANA/PTI functions through evaluating, reviewing, reporting, recommending, and updating processes. In doing so, it takes into consideration and with great care the recommendations made by another technical body (the Internet Engineer-ing Task Force or IETF) as 'Requests for Comments' (RFCs). RFCs are part of 'a consensus-based approval process' in which rank-and-file comments are welcomed, provided two principles are respected—*fairness* ('all activities are performed in a professional, fair, and neutral manner'); and gratuity ('accurate and authoritative registries are maintained and made available to the public without cost'). This architecture is meant to dilute power, depoliticize issues, and give voice to both engineers and end users. They both keep control over the whole process all along, instead of transferring their decision-making power to diplomats once preparatory committees are replaced by a plenary as in any other IO. To guarantee that ICANN does not trespass its mandate, the five Regional Internet Registries (RIRs) monitor some of its activities, while others are under the oversight of PTI (previously, IANA).[22]

The decision-making bodies operating in the field of Internet governance are contractual. They operate according to private procedures and Law, in a sort of 'checks-and-balances' system of cross-control. At the outset it was obvious that the international component of Internet Governance would go private rather than public (the US government remaining the Principal of this Agency). ICANN is not an 'intergovernmental organization', and not even a national administration since the US Department of Commerce, and neither the White House nor the Congress, was represented by the state of California.[23] It is of note that the GAC not even a primus inter pares body within the organization, is exactly on the same level as the 'At-large committee'. This means that as far as 'providing advice to the ICANN board on policy matters' is concerned, diplomats are balanced by experts (IANA 2015). Such an encroachment on diplomatic charts is reinforced by 'the accountability reform process' of the IANA transition, in which 'governments, via the GAC, were given same status as nonstate actors' operating according to the California Nonprofit Public Benefit Corporation law.

Of course, the balance of power may still change, even after the end of the transition of IANA stewardship to ICANN in October 2016, which brought an 'empowering (of) the global Internet community to have direct recourse if they disagree with decisions made by ICANN' (IANA 2016, emphasis mine), 'outside of the confines of existing multi-lateral governmental institutions' (Post and Kehl 2015, 29). The balance between communities and states is nonetheless unstable, since the GAC has 'progressively extended its influence'. While 'the GAC only provided advice when the board asked for', it is now enabled to make recommendations. After the demise of the WSIS, it even started to 'have the last word' each time its statements departed from proposals made by the Generic Names Supporting Organization. However, it seems unlikely that 'a combination of intended and unintended consequences of changes in the role of the GAC could subvert the fifth principle of the

NTIA over the longer term', and the 'Internet's idea of permissionless innovation would be thrown out the window' (Mueller 2015).

ICANN's degree of 'publicness' (Bozeman 2007), though, does not depend only on its QUANGO (quasi nongovernmental) status, or the vigilance of IGOs and governments. It is 'public' insofar as it serves the general public with no restriction: ICANN assigns to board members and Liaison officers the responsibility to act 'to the best interest' of their body, and eventually in 'the global public interest' (the 'national interest' is noteworthy absent from this statement, a great departure from classical diplomacy). It is also public as 'the administrator of a global public trustee regime' whose sovereignty competes with national governments claims over a territory, even when the issue at stake is the assignment of country codes Top-Level Domains, or ccTLDs, like '.UK' instead of '.ORG' (Mueller and Badiei 2017, 491). Hence, it does not need public agents like diplomats who respect public protocols and steer a publicly owned structure under the umbrella of a national constitution. Contracting transnationally with private individuals who are not at all related to governments (since they operate within a 'Cross-Community' framework), and relying on 'the idea that the Internet community would make its own public policies' are therefore sufficient to fulfill Internet-related functions (Teubner 2012, 57–58; Mueller 2015, 15).[24]

Such options (multistakeholderism and privateness), have long-lasting and deep consequences on the relationships between every components of the Internet governance system: decentralization brings a new model of strategic interactions, which is perfectly *devoid of mediations,* and looks rather egalitarian, contrary to the kind of centralization that is at the roots

Table 2.2. Two ideal types of Internet governance

	Multilateralism	Multistakeholderism
Network composition	Intergovernmental organization	Meta organization? Hyper social network?
Network functions	Public system of intermediation	Network of private intermediaries
Decision-making process	Exclusive (diplomats have the last word)	Inclusive (crowd-sourced)
Publicness	National public entities	Global public interest
Legal status	Constitutional (refers to national and international Law)	Contractual (unending renewal process)
Agency	Diplomats debating behind closed doors	Pseudo-diplomats, conversing in a techno democracy
Oversight	Governments representatives and IGOs staff, top down	Community of experts, cross-control

of diplomacy. Actually, the essence of diplomacy is to prioritize targets, issues, and people. There is nothing more formalized than diplomacy, indeed, with its pecking order of country representatives when they meet anywhere (from a state reception to an IO annual meeting, not to speak of negotiating tables). Sitting them in the right order of precedence is compulsory; learning this order is on the list of rules in the booklets that are given to every newcomers.

This is why the opposition between 'Internet Protocol' and 'diplomatic Protocol' is not just a pun. It addresses an actual difference between two opposed systems of regulation, as shown on Table 2.2.

CONCLUDING REMARKS

In the past, diplomats created and manipulated spatial buffers and timely pauses, which paved the way to second thoughts and opinion change—the very goals of persuasion. Time and space are in effect needed to convince people, make them abandon their initial beliefs, and even behave in contradiction with their initial pledges. The essence of classical diplomacy was to avoid tipping points when, instead of reaching a consensual agreement, stakeholders tied their hands before stepping in the negotiation room with such determination and publicity that they could not look back and make last-minute concessions.

The famous distinction between front stage and back stage does just this: whatever the means to connect the former to the latter (a curtain as in a theater; lobbies and bridges separating one hotel from its annex; streets between two blocks; and above all, encrypted transactions between negotiating spots and national headquarters), they give time to rectify a losing policy while preserving the face of the losers and they also make secret moves possible without risking naming and shaming harassment attempts (Cicourel 1988, Mansour 2011).[25]

In my view, the Internet destroys all this. Any person can at anytime an anywhere press 'enter', and immediately makes plain her position to the rest of the world. Degrees of publicity are simply leveled off. Succession of steps is just forgotten.

As for the governance structure, and contrary to what inevitably stroke any regulation system ever known in the past, the ICANN-centered cobweb has not given birth to an organizational and inclusive cluster of partners among which states and IOs would have enough leverage on an equal footing with other agents. Among such agents, heterogeneity is the rule: there is the federal administration of the United States, the state of California, American private stakeholders, and representatives from the rest of the world. Civil servants and engineers are center stage; diplomats are not.

Historically iterative discussions among private actors relayed by governmental representatives after preparatory work had been done, ended up in the making of 'treaty organizations'. This was at the roots of the existence of the WMO, the WIPO (Schemeil 2004), ILO, and IMO—to name but a few—and, of course, the whole humanitarian system headed by the UNHCR and the ICRC. On the contrary, in the realm of the Internet, governments have not succeeded in retrieving any substantial decisional power so far.

Of course, the founding fathers of the Internet and their successors have endlessly repeated that the tasks to complete were 'technical'. However, there is politicking everywhere, even in the most mechanical machinery (Petiteville 2018). High politics is never far away. This is when diplomats step in, to solve the 10 percent problems on which experts still disagree. They know by experience that arguments cannot be tabled as such, i.e., expressed bluntly, spontaneously, immediately, and publicly: motivations must be reframed by professionals, if only because compromises depend on delaying, postponing, buffering, concealing, and translating—the sort of diplomatic skills that need to be learned. Are such actions compatible with the Internet values? Can the hybridity of new philosophy and old practice bring 'techno-diplomacy' (Der Derian 1999)? Doubtful.

At the time diplomacy was invented it was meant to translate violent conflict into symbolic confrontations (Warren 1989) or generate 'different fears about the consequences of credible threats' (Trager 2017). The contention that diplomacy was thought out to 'mediate alienation' and 'estrangement'[26] still applies to Internet governance half a century after Der Derian's first paper on diplomacy was out (Der Derian 1987) and thirty-five years after his masterpiece was published (Hill 1987; Der Derian 2017).

Within this new context, though, 'face-to-face negotiations that were meant to calm down fears and, help state representatives to make credible inferences from their meetings' are getting rare (Trager 2017). This trend can deprive mediators from 'discussing alterity and the potentially destructive consequences of turning difference into otherness' (Leira 2017, 70)—a well-known bias of unmediated exchanges on the web. In previous centuries, 'representatives' acted distinctly on behalf of their territorialized governments. They were accountable to them only. They did not act on behalf of a de-territorialized community according to a fully new principle, the pre-eminence of a universal 'Sole Member Community Mechanism'. They were not accountable to the international community of the peoples (McGillivray 2014). On the contrary, within the framework of its libertarian philosophy 'ICANN should not be involved in determining who is the "best" delegee (sic) for a territory' (Mueller 2017a). This is a dramatic change from the past.

The diplomatic/organizational world have tried hard and for years to make a great comeback in this realm, as shown by the relative resilience of the ITU (Calderado 2021), the still-frustrated ambitions of the GAC, and an increasing UN interference via the WSIS, followed by the rescue of the digital divide by the IGF, plus further attempts to restore UN authority both via the ITU during the 2018 renegotiation of the 1988 ITU Regulations (ITUR); and when the Open-Ended Working Group (OEWG) of the UN was launched that same year.[27] Such effervescence gives an impression of a re-diplomatization of the sector (Mueller 2015; 2017b)—in simpler terms, a return to multilateralism in the fields of international relations where it had collapsed. However, this may be a false perception. Or is it?

Despite such indicators, one may wonder if public diplomacy in this field had not yet become private dialogue (Cull 2013; Robin 2005)? The 2016 transfer of stewardship on IANA functions from the US administration to ICANN may be the last step before a public system of intermediation becomes a network of private intermediaries. If this were true, horizontal coordination would replace vertical hierarchy. And centralized coexistence (the diplomats' century-long achievement) would be replaced by decentralized fragmentation[28] (a geeks' ambition of the millennium[29]) (Schemeil 2012). Would this occur, then any 'alignment with national sovereigns' (Mueller 2017a) could become impossible and with it any return to some rejuvenation of classical diplomacy at least on multilateral stages would be illusory.

NOTES

1. As this author puts it: 'Diplomacy can both make agreements that should be easier to reach more difficult and make agreements that should be hard to achieve more attainable'. He also states that, among diplomats, 'prosocials' (and 'value-creating') overcome 'proselfs', i.e., 'value-claiming' (Rathbun 2014: 11).

2. Note that there are popular issues about the possible connexion between 'diplomacy' and 'Internet' that I shall not address here, i.e., the use of digital technologies by diplomats; the comparative advantage of diplomats from technologically advanced countries in which new software is invented over their less advanced peers; and the future of a particular conception of democracy born in mid fifteenth century Europe. On this last point, James Der Derian recently confessed this about his first meeting with his PhD supervisor: 'I think it was the third or fourth week that (Hedley) Bull gave me the question that would eventually inspire *On Diplomacy* . . . 'Does *western* diplomacy have a viable future?' (my emphasis).

3. The actual difference between the two ages of diplomacy (before and after the Internet) may lie elsewhere—in the recent break between state and society. Classical

diplomats produced reports and joined collaborative projects; they respected the principle of non-interference in another state's affairs. New diplomats increasingly support initiatives from civil society activists (Cornut 2018, 726–29).

4. 'The diplomats themselves understood that only regular face-to-face meetings could reduce uncertainty in negotiation'. 'The outcome of a diplomat attempting to understand an interlocutor's intentions that are communicated in a letter or cable wire should be different than a diplomat attempting to understand intentions through a face-to-face interaction' (Holmes 2013, 841; 853).

5. Try this: write 'solicitude' with a misspelling and forget to check that the proofreader did not put 'solicitation' instead, quite the opposite of what you actually meant.

6. Interview, Victor do Prado, WTO Director of the Council and the Trade Negotiation Committee, March 16, 2017.

7. Even 'the UN has experimented with new kinds of less formal "ambassadors" . . . The "Goodwill Ambassador" system, which today numbers some two hundred appointees, emerged by incremental practice starting in the 1950s. The "Messengers of Peace" program came about partly to bring organizational coherence to the uncoordinated and not fully formal practice of appointing goodwill ambassadors' (Wiseman 2015, 321).

8. Science can also replace diplomats, when scientists force governments (and their permanent representatives) to protect the Commons: 'in the current multilateral organisation of world politics, any state can still refuse at any time to comply to international rules that are perceived as not serving its interests . . . This is the major issue for the global commons: who cares? The answer might be: scientists' or, better: a 'Global Science for Global Diplomacy', which 'implies that . . . scientists and diplomats need to step up their interactions in the context of a multi-lateral mode 2.0 environment' (Van Langenhove 2019, 20).

9. As put by Epstein, Roth, Baumer (2014): 'The IGF was created in 2006 out of a diplomatic impasse regarding authority over Internet governance during the World Summit on Information Society (WSIS) . . . The IGF has no negotiated outcome and has been criticized by a number of governments (most noticeably those among the BRICS group: Brazil, Russia, India, China, and South Africa) and watchdog-oriented civil society organizations'.

10. As Goffman wrote: 'Minor social ritual [that is, practice] is not an expression of structural arrangements in any simple sense; at best it is an expression advanced in regard to these arrangements. Social structures don't "determine" culturally standard displays, merely help select from the available repertoire of them. The expressions themselves, such as priority in being served, precedence through a door, centrality of seating, access to various public spaces, preferential interruption rights in talk, selection as addressed recipient, are interactional in substance and character; at best they are likely to have only *loosely coupled* relations to anything by way of social structures that might be associated with them' (emphasis mine, quoted by Pouliot (2016, 13)).

11. 'With the exception of the Liaison appointed by the Governmental Advisory Committee, any Liaison may be removed, following notice to that Liaison and to the

organization by which that Liaison was selected, by a three-fourths (3/4) majority vote of all Directors if the selecting organization fails to promptly remove that Liaison following such notice. The Board may request the Governmental Advisory Committee to consider the replacement of the Liaison appointed by that Committee if the Directors, by a three-fourths (3/4) majority vote of all Directors, determines that such an action is appropriate' (See https://www.icann.org/resources/pages/governance/guidelines-en).

12. See ICANN website, Governance: https://www.icann.org/resources/pages/governance/guidelines-en.

13. 'The Ombudsman shall serve as an objective advocate for fairness, and shall seek to evaluate and where possible resolve complaints about unfair or inappropriate treatment by ICANN staff, the Board, or ICANN constituent bodies, clarifying the issues and using conflict resolution tools such as negotiation, facilitation and "shuttle diplomacy" to achieve these results' (See Charter section 5.2 https://www.icann.org/resources/pages/governance/bylaws-en#V).

14. See ICANN's Organization chart available at https://www.icann.org/resources/pages/chart-2012-02-11-en

15. 'ICANN's role is to oversee the huge and complex interconnected network of unique identifiers that allow computers on the Internet to find one another. This is commonly termed 'universal resolvability' and means that wherever you are on the network—and hence the world—that you receive the same predictable results when you access the network. Without this, you could end up with an Internet that worked entirely differently depending on your location on the globe.' (See https://www.icann.org/resources/pages/what-2012-02-25-en).

16. See https://www.icann.org/resources/pages/what-2012-02-25-en, emphasis mine.

17. See ICANN website: Guidelines, 12, 23, 30, 32; https://www.icann.org/resources/pages/guidelines-2012-05-15-en.

18. 'Meta-organizations are autonomous actors having autonomous actors as their members. This paradox has numerous ramifications for the way IGOs are established, how they make decisions, and how they can influence their members and the wider environment. Overall, Meta-Organization Theory . . . explains why IGOs are weaker actors than standard organization theories would suggest, and they gain their influence in a different way' (Ahrne, Brunsson and Kerwer 2016).

19. As board chair Stephen D. Crocker said in October 2016: 'This community validated the multistakeholder model of Internet governance. It has shown that a governance model defined by the inclusion of all voices, including business, academics, technical experts, civil society, governments and many others is the best way to assure that the Internet of tomorrow remains as free, open and accessible as the Internet of today' (IANA 2016).

20. 'At the end of that process (once any of the supporting organizations has completed a draft), the ICANN Board is provided with a report outlining all the previous discussions and with a list of recommendations. The Board then discusses the matter and either approves the changes, approves some and rejects others, rejects all of them, or sends the issue back down to one of the supporting organisations to review, *often*

with an explanation as to what the problems are that need to be resolved before it can be approved.'

21. 'Today the services are provided by Public Technical Identifiers, a purpose-built organization for providing the IANA functions to the community. PTI is an affiliate of ICANN.' (https://www.iana.org/about).

22. Before the transition, the latter was itself controlled by NTIA (the United States Department of Commerce's National Telecommunication and Information Administration), and not by other governments packed into the GAC: 'NTIA also verifies that ICANN followed established process in requesting changes to the authoritative root zone' (IANA 2015). On the IANA transition see Palladino and Santaniello (2021).

23. According to Mueller (2015), 'the founders of ICANN' created it 'to boost support for the regime among other governments'. More information on the GAC can be found on its dedicated website: https://gacweb.icann.org/display/gacweb/Govern mental+Advisory+Committee. More on the At Large Advisory Committee: https://atlarge.icann.org/alac.

24. Following on Renner's 'private global regimes' (Renner 2010), Guenter Teubner infers that ICANN is an example in '"private ordering" from the existence of transnational Internet private arbitral tribunals', which do not refer to national constitutions but to the universal principle of freedom of opinion (Teubner 2011, 195–96; 2012, 57–58). As a 'global Law enforcer', ICANN designed a 'Public Interest Commitment Dispute Resolution Procedure' or 'PICDRP' (Post and Kehl 2015). Arbitrage centers are the WIPO Arbitration and Mediation Center, the National Arbitration Forum, eResolution, and the CPR Institute for Dispute Resolution; but there also are national or regional Dispute-Resolution, and Service Providers such as the Kuala Lumpur Regional Centre for Arbitration or KLRCA, see https://klrca.org/. 'As a dispute resolution service provider accredited by Internet Corporation for Assigned Names and Numbers (ICANN), KLRCA administers domain name dispute resolution proceedings. The KLRCA has also been appointed the .my domain name dispute resolution service provider by the Malaysian Network Information Centre (MYNIC), which administers the .my domain'.

25. This was why Kennedy opted for a blockade of Cuba far away from its coasts in 1962: he wanted to buy time for diplomacy and stop the race to the bottom.

26. As James Der Derian (2017) puts it: 'Without getting too personal, I also suspect that those drawn to the art and practice of diplomacy have in common familial histories of estrangement that engendered mediatory practices of one kind or another'.

27. 'Hosted by the UN, representatives of the multistakeholder community of the OEWG regularly met to negotiate principles of international cooperation towards an "open, secure, stable, accessible and peaceful ICT environment", a process finalized in March 2021 with the release of the "UN OEWG Final Substantive Report"' (Calderaro 2021).

28. Fragmentation is not understood here as in Milton Mueller's work, since he defines it as 'segmentation of cyberspace into *national* jurisdictional spaces' whereas in this chapter we are addressing the segmentation of *transnational* spaces into multiple cross-controlling bodies. Furthermore, Mueller goes much further than he usual Internet stakeholders when they criticize 'the dangers of enhanced governmental

influence' and promote a 'transnational popular sovereignty' (Mueller 2017a, my emphasis; 2015).

29. At the very least it must have been John Postel's dream, since he ruled the first Internet informally from his lab in the University of Southern California before the NTIA started to control IANA (i.e. when 'IANA's exact legal authority was difficult to discern'). In such an informal and improvised context he 'went as far as to make direct claims of authority over the DNS root' (McGillivray 2014, 7).

Chapter Three

Diplomacy and Internet Governance

A Conceptual Re-Assessment

Katharina E. Höne

What is the relationship between internet governance (IG) and diplomatic practice? To what extent is diplomatic practice driving IG by shaping processes of deliberation and decision-making? Conversely, are developments in the area of IG—such as new technologies but also new forms of deliberation and governance—shaping diplomatic practice? Posing these questions challenges us to rethink the practice of diplomacy and the impact of IG. As a social practice, diplomacy is never static. Rather, it adapts to a changing environment, a changing set of participants, and the emergence of new topics. In addition, questions like these also prompt us to carefully identify and critically engage with the actors and processes of deliberation that shape IG.

This chapter argues that the relationship between IG and the practice of diplomacy is a complex one that is characterized by mutual influence. In order to sustain this argument, the chapter is interested in both conceptual as well as practice-oriented reflections. As such, the chapter makes four moves. It begins with a general discussion of the connections between IG and diplomacy; addresses the conceptual relationship between IG, in particular questions of global governance, and diplomacy; looks at the role of new actors, which are enabled by technological innovation and new practices related to IG; and discusses technology as shaping but also as being shaped by diplomatic practices. In mapping the relationship between IG and diplomacy, this chapter argues that diplomacy is the process driving IG. Yet IG also provides new topics, tools, and technologies that question and reshape established diplomatic processes. The chapter discusses the literatures on conceptions of diplomacy, global governance, and IG and relates them to current practices.

IG AND DIPLOMACY:
INTERCONNECTIONS AND DICHOTOMIES

This chapter addresses the relationship between IG and diplomacy from a conceptual as well as practice-oriented perspective, while taking the impact of technology in enabling and shaping diplomacy, and conversely the importance of diplomacy for shaping the international policy environment for new technology and IG, seriously.

If we are to think about IG through the lens of diplomacy, it is crucial to acknowledge that some aspects of IG go directly against the cornerstones of a traditional understanding of diplomacy. IG raises questions about territorial conceptions of sovereignty and the primacy of the state and its representatives in managing international relations—core assumptions of diplomacy traditionally understood. The internet as a technology has not only raised new questions; it has also enabled the participation of new actors, provided new tools for communication, and made state borders more permeable.

At the same time, diplomatic processes and their outcomes have been shaping IG. For example, it is worth noting that the term IG, as we use it today, is the result of a diplomatic process, the World Summit on the Information Society (WSIS), held in 2003 and 2005. The definition agreed at WSIS describes IG as 'the development and application by governments, the private sector and civil society, in their respective roles, of shared principles, norms, rules, decision-making procedures, and programs that shape the evolution and use of the internet' (ITU 2005). The term itself appeared for the first time in internet related discussion during the 1980s. It, however, only entered the negotiations leading up to the 2003 WSIS at a relatively late stage (Kurbalija 2017, 7–8). So, while the practice of governing the internet was well underway before the term was coined, the WSIS process brought about an agreed-upon definition.

Yet, despite the fact that WSIS provided a definition of IG, the meaning of the term is not at all fixed. Rather, 'it varies according to the background and objectives of those who invoke it' (Brousseau and Marzouki 2012, 368). The actors, topics, aims, and processes of IG discussion remain in flux and can only be pinned down temporarily (Radu 2019, 9–11). They include traditional diplomatic actors as well as new ones. This is both a strength and a weakness of IG debates and processes. On one hand, it remains open to new issues and new actors entering the debate. On the other hand, vagueness can undermine efforts towards institution building and addressing specific policy issues.

Similarly, we need to be aware that early activities in IG—taking place before 1994—were managed by a small technical and academic community.

The desire to keep the state and its representatives out of IG discussions is well reflected in J. P. Barlow's 1996 *Declaration of the Independence of Cyberspace*, which states that '[the internet] is inherently extra-national, inherently anti-sovereign and your [states'] sovereignty cannot apply to us. We've got to figure things out ourselves'. As a technology, the internet has retained this potential to undercut established power structures.

The interconnections and dichotomies between IG and diplomacy briefly outlined here are further explored in the following sections, which focus on the relationship between diplomacy and global governance in the context of IG, non-state actors in the context of new diplomacy and multistakeholderism, and new topics and tools for diplomacy in the context of emerging technologies, respectively.

IG BETWEEN DIPLOMACY AND GLOBAL GOVERNANCE: CONCEPTUAL REFLECTIONS WITH EMPIRICAL AND NORMATIVE IMPLICATIONS

IG carries governance in the title, and it does indeed have a close connection to global governance as a concept. This raises the question: where does diplomacy fit? Hence, it is necessary to discuss the relation between diplomacy and global governance on a conceptual level and to relate this discussion to practices in IG. This section pursues the argument that we do best when we locate IG between global governance and diplomacy and concludes with a concrete suggestion for discussing IG in terms of both.

In conceptual terms, the governance part of IG can be a source of confusion. For example, the fact that governance is not the same as government, though a commonly accepted point, needs to be repeated frequently (Kurbalija 2017, 6). Similarly, this also raises questions about the relationship between IG and the global governance debates. How neatly does IG fit into the global governance movement?

The term global governance emerged in the 1970s and gained full prominence during the 1990s. In a sense, it was intended as a challenge to the dominance of the state and power politics in international relations and to provide a way of expressing a desire for changes in the way in which international relations were conducted. In other words, it is employed to capture what traditional conceptions of international relations, and by extension diplomacy, with their focus on the state could not. The processes that global governance aims to capture take place below, above, around, and with the state (Badie 2013, 86). One of the most prominent definitions is the one put forward by the Commission on Global Governance in its *Our Global Neighborhood* report, which emphasizes that both individuals and institutions, in private and public functions, take part in global governance in order to manage their common affairs. It includes formal institutions as well as informal arrangements that express common rules and accepted norms (Commission on Global Governance 1995, chapter one).

Two developments in particular influenced those thinking about global governance. First, with the end of the Cold War, new opportunities for international cooperation emerged and, associated with this, a new enthusiasm about the possibility of international cooperation. Second, globalization and market liberalization raised new questions about the role of the state and the best ways of addressing these processes.

It is important to note that global governance is both an empirical and a normative challenge to traditional international relations and diplomatic studies. Empirically, traditional conceptions of diplomacy had a hard time capturing the emergence of international regimes and new actors in international relations. Normatively, traditional notions in diplomatic studies and international relations also appeared to stand in the way of changes towards greater inclusivity and mechanisms that could address emerging global problems more effectively. As such, descriptions of global governance as a 'pattern of transparent and inclusive processes to address complex transnational collective-action problems' build on empirical observation but also make a claim for the normative value in fostering such patterns (Cooper, Hocking, and Maley 2008, 1). Unsurprisingly, we find substantial overlap between global governance and IG, and the 'governance' in IG is conceptually related to many global governance approaches (Mueller, Mathiason, and Klein 2007).[1]

This understanding of global governance needs to be contrasted with conceptual reflections on diplomacy. As alluded to in the previous section, diplomacy, in its modern form, is very closely intertwined with the sovereign state and hence also territoriality and the task of representing the state. Those engaged in diplomatic studies tend to define the practice of diplomacy as 'the dialogue between states' (Pigman 2010, 5) and as a specialized activity carried out by selected officials who enjoy special privileges and immunities. Harold Nicholson, the famous early-twentieth-century diplomat and scholar, described diplomacy as 'the management of relations between independent States by the process of negotiation.' (as quoted in Freeman 1993, 101) Its purpose is to enable states to secure the objectives of their foreign policies without resort to force, propaganda, or law (Berridge 2015, 1). It achieves this mainly by communication between professional diplomatic agents and other officials designed to secure agreements (ibid.). These definitions reflect a traditional conception of diplomacy and highlight its links with territoriality and sovereignty. The emphasis on the primacy of the state and its representatives, or what can be called 'the diplomacy of states' (Sharp 2019), is also noteworthy.

Contrasting these understandings of global governance and diplomacy, we see a clear dichotomy and tension between the two (Cooper, Hocking, and Maley 2008, 2). In fact, there is a tendency for scholars to either work within the conceptual framework of diplomacy or the conceptual framework of global governance; few works use the terms together or even try to establish a conceptual relation between them (exceptions include Cooper, Hocking, and Maley 2008; Fidler 2011).[2]

The differences seem obvious. Where traditional definitions of diplomacy portray it as an activity exclusive to a selected group of individuals, global governance emphasizes the participation of all stakeholders. Where diplomacy emphasizes the primacy of the state and focuses on interstate relations, global governance questions, and sometimes aims to subvert, the dominance of the state and its representatives, and takes various levels of activities, above and below the state, into account. As such, the dichotomy between diplomacy and global governance becomes a description of opposing worldviews regarding international relations and the role of the state and globalization.

In this understanding, whether one uses the term governance or the term diplomacy comes down to the question of (normative) emphasis or empirical fit. In other words, the tendency is to reserve the term diplomacy for policy fields dominated by state-actors, such as disarmament and national security, and to employ the term global governance to areas of international relations that show a greater involvement of non-state actors, such as digital policy. Similarly, those that wish to make a normative claim for the need to involve non-state actors more fully in global agenda-setting and policy-making, and to add greater transparency to global decision-making, will tend towards operating within the global governance framework.

If we stop our elaborations on IG and diplomacy here, we are faced with a conundrum. Bringing diplomacy to bear on IG encourages the charge of either trying to bring the state back in with full force or of making the claim that state actors should indeed be the dominant actors in the IG debate, while neglecting the importance of other stakeholders. Similarly, if we adopt an empirical view, diplomacy in IG appears to be something only relevant for those aspects and arenas of IG in which states can be seen to take a more pronounced interest at any given moment. Neither of these options is desirable.

Some observers have identified a growing convergence between diplomacy and global governance on both empirical and norm-related grounds. They identify an opening up of diplomacy to non-state actors and a greater pragmatism regarding global governance projects (Cooper, Hocking, and Maley 2008, 4). However, whether or not the convergence is conceptual or indeed also empirical, it harbors the fear that diplomacy is 'losing both its professional and conceptual identity' (Sharp 1997, 630).

In response to the conundrum, I propose to focus our understanding of diplomacy and global governance on what each has to offer as their main analytical strength. In a first step, we need to return to a definition of diplomacy that goes beyond its Westphalian moment to look at diplomacy as a practice that is not necessarily tied to the state. As such, it is useful to look at the functions of diplomacy as laid out, for example, in the Vienna Convention on Diplomatic Relations (VCDR, article 3): information gathering, communication, representation, negotiation, and the promotion of friendly relations (compare also Neumann 2008; Wight and Butterfield 1969). Focusing on the functions of diplomacy allows us to look at the kinds of actors that fulfill these functions in a given context and therefore loosens the focus on the state. Thus, we can reconceptualize diplomacy as a specific activity and focus on the relevant actors and the practices they are engaged in.

In a second step, we can look at the main analytical strength of the global governance literature. Global governance clearly stands out for its emphasis on rules and norms and their institutionalization. Regime building and the emergence of international structures are well captured by global governance approaches and, at the same time, often escape the purview of those studying diplomacy.[3]

The suggestion made here is to reserve the term diplomacy for actors and the processes between them and to use the term governance to describe informal and formal institutions and rules of behavior (compare also Fidler 2011). In this framework, diplomacy describes various practices, most important among them the articulation of interests and the process of bringing these interests to negotiated solutions. Where these interests converge sufficiently over time, they can lead to formal as well as informal institutions. In other words, governance structures and rules of conduct emerge. In turn, these governance structures and rules of conduct shape the practice of diplomacy. Yet, it is diplomatic practices, carried out by a variety of state and non-state actors that work to re-produce but also change these governance structures. Ultimately, such a distinction brings us back to the heart of social theory, for example Giddens's structuration theory, and the relation between structure and agency (Giddens 1984).[4]

Further, it not only enables us to think diplomacy and global governance together but to also address the shortcomings in diplomatic studies, with its tendency to shy away from the analysis of institutions and non-state actors, and the shortcomings of the global governance literature, which, at least in its institutional liberal guise, has a tendency to focus on institutions to the detriment of processes and actors. Hence, it is also useful to discuss IG in terms of diplomacy.

As we have seen, the traditional definition of diplomacy fails to capture some of the important actors involved in IG. The term 'new diplomacy' has been coined to capture the emergence of new diplomatic actors and so-called multistakeholderism addresses their involvement in multilateral diplomacy. Both of these new developments in diplomacy are discussed in relation to IG in the following section.

NEW DIPLOMATIC ACTORS—REFLECTIONS ON 'NEW DIPLOMACY' AND MULTISTAKEHOLDERISM

As we have seen, a large part of diplomatic studies, like many aspects of scholarship on international relations, still focuses on the state as the main actor and gives it an elevated position. This position has come under critique, which builds to a large extent on empirical observations about the emergence of new actors in arenas traditionally reserved for representatives of states.

This position contrasts traditional diplomacy with so-called 'new diplomacy'. The term describes the appearance and growing relevance of new actors and new topics on the global agenda. It seeks to capture the fact that international relations are no longer exclusively conducted by states and their representatives. Many of the debates about new diplomacy emerged from global environmental negotiations, such as the 1992 Rio Earth Summit, and, more recently, have found new impetus through the Sustainable Development Goals (SDGs) and the process that led to their adoption. Kjellén, chief negotiator for Sweden's environment and climate change diplomacy in the 1990s observed that 'we have entered a new era of international cooperation [in which] the boundaries of traditional diplomacy—concentrated on national security and economic and commercial matters—are being extended to a much broader concern for global sustainability' (Kjellén 2008, xvi). Similarly, the consultation process leading to the 2015 adoption of the SDGs has been praised as exceptionally inclusive and open to non-state actors.

In meaning, new diplomacy is closely intertwined with multistakeholderism. Put simply, multistakeholderism aims for the involvement of (all) relevant stakeholders, which can be understood as 'groups of actors organized around specific common principles, values, visions, legal status, and organizational structures and that have a certain stake in a process or issue' (Kleinwächter 2008, 537). It is a term most closely associated with the WSIS process and it, hence, has gained particular importance for IG.

Yet, it is useful to keep in mind that proclamations of a new diplomacy are, in fact, not new. An example of a practice termed 'new diplomacy' is US president Wilson's promotion of multilateral and parliamentary-style diplomacy in the inter-war period of the 20th century (Kelley 2013). Much like multistakeholderism today, these ideas were associated with hopes for fundamental reform of the conduct of international relations (Wiseman 2011, 8). Critics at the time, much like critics today, feared that this would amount to nothing more than dressing up established institutions in new clothing. A formidable concern is raised by Berridge, who argues that the term new diplomacy is simply employed to mark 'the latest fashion in diplomatic method' (Berridge and Lloyd 2012, 259). From this perspective, new diplomacy and multistakeholderism serve as a façade behind which the re-emergence of traditional state power and secret negotiations hide. Berridge identifies this as the 'counter-revolution in diplomacy' (Berridge 2011, 1–15). While we can indeed observe the emergence of new actors in relation to fora traditionally reserved for the state and its representatives, critics argue that these new actors do not have the same decision-making power. The concern is that new diplomacy and multistakeholderism are a terminological smokescreen, which serves to hide the fact that older power relationships, centred on the state, are not only still alive but also dominant in the conduct of international relations. This point is important to stress and has some bearing on IG processes and the role of state representatives.

IG is very much a reflection of both the hopes associated with new diplomacy and the push-back from traditional actors. As mentioned in the first section of this chapter, the early days of IG were marked by the prominence of non-state actors in discussions about the regulation of the internet. IG was perceived as a technical issue, and the technical community claimed authority while emphasizing self-regulation (Radu and Chenou 2014, 4). Some participants in this debate elevated this absence of the state to a normative guiding principle, as for example expressed in Barlow's *Declaration of the Independence of Cyberspace*. This changed with the so-called Domain Name System (DNS) wars, which were essentially disputes over the management of the key infrastructure of the internet and led more traditional diplomatic actors—states and international organizations—to enter the scene (Kurbalija 2017, 7). While this is perhaps unsurprising, it is an important counter to the hopes associated with new diplomacy and multistakeholderism. As the political and economic stakes in the regulation of the internet shifted, so did the balance between state and non-state actors as well as between various non-state actor groups in the debate. Despite this shift, which came about mainly due to US authorities being more visible and involved, the importance of including a variety of stakeholders was generally recognized. In 1998 for example, US authorities argued that the legitimacy of the Internet Corporation for Assigned Names and Numbers (ICANN) is tied to the inclusion of key stakeholders. Yet, we also need to acknowledge that the very process that led to the establishment of ICANN left some stakeholders marginalized (Radu and Chenou 2014, 5 and 9). In other words, in the area of IG, the new diplomacy had to face its own 'counter-revolution'.

Despite this, IG is a good example of a space in which the inclusion of non-state actors is driven by functional necessity. Generally speaking, the highly technical nature of some aspects of IG makes the inclusion of non-state actors who bring this specialized expertise to the table crucial. It is safe to say that it was more this practical concern, rather than a moral commitment to inclusivity, that led to multistakeholderism becoming an established part of the process at the 2003 and 2005 WSIS summits.

While the multistakeholder model is firmly intertwined with the WSIS summits and enshrined in many IG processes we need to wonder to what extent the inclusion of all stakeholders also means that these stakeholders have a say in the decisions that determine the future direction of IG. The transition of the Internet Assigned Numbers Authority (IANA) functions reminds us that multistakeholderism cannot be taken at face value (Hill 2016). In other words, multistakeholderism needs careful examination, especially with regard to the extent to which various stakeholders' views are able to influence negotiations, policies, and decisions. The 2003 Geneva WSIS summit itself is exemplary of this dilemma. While all stakeholders gave substantive input during the pre-negotiations, the so-called prep-comms, the moment at which the negotiations moved to the final intergovernmental negotiations, non-state actors 'had no voice or vote anymore' (Kleinwächter 2008, 560). While the involvement of civil society in the early stages of the process shaped some of the content of the Geneva Outcome Documents, the documents themselves were agreed though an intergovernmental process.

Other IG-related examples that follow the multistakeholder model are the Working Group on Internet Governance (WGIG), the Internet Governance Forum (IGF), and NETmundial initiative. WGIG, formed after the 2003 Geneva summit in order to give input for the 2005 WSIS in Tunis, was 'as a specific diplomatic mechanism, based on equal participation of governments, civil society, and the business sector' and can be seen as 'an innovation in the existing, mainly inter-governmental, diplomatic system' (Kurbalija 2008, p. 180). The IGF takes the form of a policy dialogue, which brings together all relevant stakeholders in an annual meeting. While in principle adhering to the multistakeholder model, the IGF is criticized for the lack of tangible outcomes (Doria 2014) and there have been calls for improved cooperation, taking, for example, the form of the so-called

IGF Plus model (UN 2019; UNGA 2020). NETmundial (2014–2016), also known as the Global Multistakeholder Meeting on the Future of Internet Governance, was an effort to create an alternative mechanism for IG with a strong focus on multistakeholderism, which began in 2014 and ended in 2016. The backing of the US and Brazilian governments promised weight behind the initiative. The proposed governing structure, however, raised controversy with key players like the Internet Society and many other civil society organizations. The initiative failed, but it can be argued that the discussion surrounding it led to 'a more open and much needed exchange about power mechanisms and imbalances' in IG (Pohle 2015). At its core, it exemplified the debate between two different models of governing the internet that are cornerstones of any future debate: multistakeholder and intergovernmental approaches.

The IGF and the NETmundial initiative remind us of some of the questions that multistakeholderism raises: to what extent are non-state actors not only at the table but also involved in shaping and taking decisions? To what extent are there attempts by traditional diplomatic actors to assert their primacy and to guard their ultimate decision-making power?

Nevertheless, it is fair to say that IG is one of the best examples we have of new diplomacy. As such, when we speak of diplomatic actors in IG, we should include state and non-state actors alike. Yet, IG is also a reminder that terms such as new diplomacy and multistakeholderism need careful examination and that 'the counter-revolution' is always just around the corner.

As we have seen, new diplomacy is often defined through the inclusion of new actors and new topics on the international agenda. In the case of IG, both are in part enabled and driven by technological innovation. This means that new tools in the practice of diplomacy, as brought about by IG-related technological innovation, also need to be acknowledged. The next section will focus on these new topics and tools in diplomacy.

NEW TOPICS AND TOOLS EMERGING AT THE INTERPLAY BETWEEN DIPLOMACY AND IG

Simply put, without such pivotal developments as the 'invention' of the web by Tim Berners-Lee, the realm of activities and topics that IG deals with would not exist. IG is driven by technological innovation that adds to and alters the topics diplomatic actors need to address and the tools available to them to do so. These tools are obviously not limited to IG-related diplomacy, but it is here that they emerge and the norms and rules regarding their use are deliberated.

The fact that diplomatic actors in IG have to deal with a number of new and sometimes highly technical topics is unsurprising. Many of the fields that fall under the term "new diplomacy," such as environment, health, or education are characterized by the fact that they emerged as topics for international relations and diplomacy only relatively recently and that they require a certain extent of specialized knowledge. IG is not an exception among these new diplomacies. Broadly speaking, the issues contained in the WSIS action lines are a good indication of the breadth of topics that need to be addressed as part of IG. A useful taxonomy is provided by arranging these into seven 'baskets' of IG comprising of topics related to infrastructure, security, and human rights as well as legal, economic, development, and socio-cultural issues (Kurbalija 2017). What is important to acknowledge is the fact that fora like WSIS and the IGF address these issues on a global level.

But more than the realization that we live in a globalized world, it is the internet as a technology that has the potential to subvert the very notion of a world neatly divisible into states with their respective jurisdictions and sovereign territories. The internet is, in part, a *disruptive technology* and this also has consequences for the practice of diplomacy (for the term compare Christensen 1997; for a critique see Danneels 2004).

Indeed, if we look at the practice of diplomacy, it is the impact of technology that is perhaps the most interesting change brought about by the field of IG. The question of how Information and Communication Technology (ICT) related to the internet would change the practice of diplomacy has been asked prominently since the mid-1990s. At the time, the influence of new ICT was very much an area for speculation. In 1996, Kurbalija for examples suggested that 'the wider use of IT within Ministries of Foreign Affairs . . . should produce in the near future a new IT-influenced diplomatic interaction symbolized by the exchange of communication through dedicated e-mail' (Kurbalija 1996, 37). While the mid-1990s were a time in which the impact of, for example, e-mail communication was very much the subject of speculation, changes in the practice of diplomacy can now be more clearly identified in the communication in and among ministries of foreign affairs, the e-drafting of documents (Adler-Nissen and Drieschova 2019), the use of mailing lists, the availability of e-transcripts of meetings and their impact on diplomatic reporting (Kurbalija 2008), the use of websites and social media (Bjola and Holmes 2015), and the use of big data and AI for diplomatic practice (Rosen Jacobson, Höne, and Kurbalija 2018; Höne et al. 2019).

Generally speaking, innovations in ICT go to the heart of the practice of diplomacy since all its functions—information gathering, communication, representation, negotiation, and the promotion of friendly relations—can be facilitated by ICT.

Having said that, thinking about changes in diplomacy in relation to innovation in ICT is not a new issue. In fact, in historic perspective we can identify an uneasy relationship between some practitioners and scholars of diplomacy and innovations in ICT. Famously, the British Prime Minister Lord Palmerston is said to have exclaimed 'my God, this is the end of diplomacy' as a reaction to receiving the first telegraph message in the 1860s (as quoted in Kurbalija 2013, 141). This fear of the demise of diplomacy as a profession was related to the assumption that better communication links would, first, make diplomats posted abroad more dependent on instructions, and second, would eventually make them obsolete as their presence in faraway regions to gather information and to negotiate on behalf of their governments was no longer needed. There is a cyclical nature to this: every innovation in ICT triggered a reaction similar to the one outlined above. Diplomacy as a profession has so far survived each of these crisis moments and has proven to be able to adapt successfully, if reluctantly sometimes.

With regard the practice of diplomacy, the internet has not fulfilled its potential or threat, depending on the perspective, as a disruptive technology. Rather, it serves as a sustaining technology, one that improves a practice (Christensen 1997). Yet, the changes are noticeable. The tools outlined above have both opened up new possibilities for the conduct of diplomacy, but also added new constrains and pressures related to the speed of communication, the vastness of information, and the pressure to adapt to new tools. Generally speaking, diplomacy has gradually adapted to new ICT while maintaining the focus on its core functions.

Nevertheless, ICT tools also had structural effects regarding the ways in which various actors can participate in the diplomatic process and relate to each other. The internet as a technology is often invoked as the driver behind leveling the playing field between state and non-state actors. It is clear that the changes we addressed as part of new diplomacy in the previous section would not be possible in the absence of ICT innovation. However, while ICT can enable more egalitarian access to information and communication, this does not result automatically in more egalitarian participation in decision-making. The norms and rules related to diplomatic participation are still subject to complex negotiation processes, which are dominated by traditional actors. To capture these norms and rules, in other words the global governance aspects of IG, a careful analysis of the diplomatic process is key.

In summary, we can see that new topics and tools in diplomacy, especially those supported by developments in ICT, mark innovation in diplomacy, however, innovation should not be confused with disruption and complete transformation (Berridge 2015, 267).[5] If the internet has disruptive potential, it has not been realized (yet) in the conduct of international relations. In part, it is the very activity of diplomacy that contributed to reigning in some of the disruptive potential by developing governance structures for the internet. In this sense, IG-related diplomacy has important implications for the general conduct of diplomacy as it is at the core of developing and reproducing the structures that govern the use of ICT.

CONCLUSION: FROM DIPLOMACY TO GOVERNANCE AND BACK

The chapter discussed the relationship between IG and diplomacy. It began by shedding light on the conceptual puzzle that we encounter when juxtaposing the terms global governance and diplomacy in the context of IG. In order to move forward productively, the chapter suggested using the term governance to refer to a particular set of rules and norms and the associated institutions. In contrast, the term diplomacy should be employed to refer to actors and their practices. In this way, we can conceptualize the relation between diplomacy and governance, including IG, without collapsing both terms into the same meaning.

Underlying this is a theoretical conception that takes the role and interplay between structure and agents seriously. Looking at the relationship between diplomacy and IG in this way highlights how the practice of diplomacy is not only changing due to the influence of ICT innovation and the demands of IG as a policy field, but is actively involved in shaping and bringing about the very conditions of this change. In other words, diplomacy shapes the field of IG, but is also shaped by it.

In order to consolidate this point, the chapter looked at the new kinds of actors that IG-related diplomacy has brought to prominence. The WSIS definition of IG reminded us that participation in IG goes beyond those actors traditionally understood to be part of diplomacy. Beyond the definition, the very process of the 2003 and 2005 WSIS summits exemplify this. In fact, many accounts describe WSIS as the birthplace of multistakeholderism—the idea that global diplomatic processes should include all relevant stakeholders. The practice of multistakeholderism raises questions regarding traditional conceptions of diplomacy. This was discussed as part of a larger trend towards what some observers have termed new diplomacy. While we can observe that the multistakeholder model, which is at the core of IG, has also led to some changes in diplomatic practices, we need to wonder to what extent this has transformed the classical model of conducting international relations driven by the dominance of the nation-state. *Plus ça change, plus c'est la même chose*? While IG is a paradigmatic example of new diplomacy, the chapter also highlighted that the 'counter-revolution in diplomacy', understood as the re-assertion of traditional diplomatic actors and forms of deliberation, is always a possibility.

Technological developments shift debates, introduce new topics, and bring about new areas in need of governance. In this sense, IG brings about new topics that need to be addressed by diplomatic practices. In another sense, IG provides new tools for diplomatic practices. At the core of diplomacy is communication; developments in ICT therefore also impact the way diplomacy is practiced (Höne 2020; Kurbalija and Höne 2021). This conceptual distinction of IG as providing topics as well as tools for diplomacy was taken up in the last part of this chapter. Diplomacy has proven to be adaptive in this regard. New ICT brings about innovation in diplomacy, however, innovation should not be confused with disruption and complete transformation. At the same time, diplomacy is the practice with the ability to reign in the more disruptive potential of ICT. In this sense, the section highlighted the importance of bringing technology into the field of diplomatic studies and the need to seriously think about the relationship between technological innovations and the practice of international relations (Kaltofen, Carr, and Acuto 2019).

In summary, the chapter provided a conceptual and practical reflection on the relationship between IG and diplomacy. It argued that diplomatic processes are at the core of shaping governance structures of IG, which in turn shape the conditions under which diplomacy, both IG-related and more general, takes place.

NOTES

1. Some scholars, however, warn against overstating the role of non-state actors in both theories of global governance and the analysis of IG debates and issue a call for refocusing scholarship on the role played by state actors. Drezner (2004) for example, argues that global governance scholarship tends to overstate the role of non-state actors. He then, however, argues for a more nuanced analysis, which transcends the binary of state versus non-state actor.

2. Fidler (2011, 6–7) develops a typology that includes both diplomacy and global governance and highlights their interconnections. He suggests to reserve the term diplomacy for process of negotiation and governance for the institutions and collective action solutions that are the result of diplomatic practice. In contrast, Cooper, Hocking, & Maley (2008) do not try to synthesise global governance and diplomacy. They argue that the two are not only distinct but also stand in opposition.

3. This, of course, comes with a caveat. The study of diplomacy is not entirely blind towards international institution. The work of scholars of the English school is just one example (Wight and Butterfield 1969). Similarly, narrowing global governance to the study of international institutions and regimes means to neglect other

aspects of global governance approaches that either focus on the relevant actors or provide an ontological criticism of both actors and institutions (Acuto 2013, 19–20).

4. In this context, it is worth noting that I treat actors and agency synonymously here. I do so largely for convenience of communication. I am, however, aware of the challenges to such a treatment and the theoretical complications entailed (Wight 2006, 178 and 190).

5. In contrast, Hamilton and Langhorne (1995) speak more optimistically of transformation and even transcendence of the traditional methods of diplomacy in light of ICT innovations.

Chapter Four

Discourse Coalitions in Internet Governance

Shaping Global Policy by Narratives and Definitions

Mauro Santaniello and Nicola Palladino

DEFINITIONS OF INTERNET GOVERNANCE AS REGULATIVE RESOURCES

The more the Internet penetrates contemporary societies, the more Internet governance emerges as an important and controversial global policy domain. Within this domain, a crucial cluster of issues is the object of multi-layered and networked processes of policymaking, which occur in different institutional settings and involve a messy galaxy of actors, actants, and hybrids. This huge complexity challenges the effectiveness of policymaking at the domestic and international levels as well, and it is reflected by the lack of a consensual and well-defined idea about what Internet governance is. As reported by Kleinwachter (2007, 14), in a 2004 speech the UN Secretary-General Kofi Annan has depicted this problematic situation as follows: 'The issues are numerous and complex. Even the definition of what is meant by Internet governance is a subject of debate'. This uncertainty about the epistemic status of IG and its basic concepts and definitions is also well recognized in literature. For example, describing IG as 'herding Schrödinger's cats', MacLean has underlined that: 'There appears to be a lack of clarity and common understanding . . . on the meaning of the terms "Internet" and "governance", when used either alone or in combination.' (MacLean 2004, 3). Similarly, Drake has highlighted that 'there is a total lack of consensus about how to define Internet governance, and about which issues and institutions are and should be involved in what manner' (Drake 2004, 1). Also, Hofmann has underlined that: 'Although the term Internet governance has been in use about ten years now, there is not yet general consensus on its meaning. First of all, the concept 'governance' and its relationship to government is unclear; second, it is unclear what extent and form of authority Internet governance has and in future should have' (Hofmann 2007, 75). Kurbalija (2014, 5) has pointed out that 'the controversy surrounding Internet governance starts with its definition. It's not merely linguistic pedantry. The way the Internet is defined reflects different perspectives, approaches, and policy interests'.

The thesis underlying this chapter is that these definitional uncertainties and ambiguous concepts are not due to the inability of scholars and policymakers in grasping the essential contours of an emerging policy field, nor to the fast pace of technological innovations. Rather, they depend on the highly controversial nature of the issues at stake in Internet governance. Since the early 1990s different actors have strategically been producing different definitions of Internet governance, in order to mark off sets of legitimated issues, actors, and fora. In other words, definitions have been used as regulative resources activated by actors in their struggle for the governance of the Internet. As noted by Hurwitz (2007, 2) 'controlling Internet governance requires controlling the conversation about Internet governance', and Saldias (2012, 3) has supported this point of view by stating that 'the conversation about Internet might even have a greater influence in controlling the Internet than formal law-making process that are currently available in international public law or national constitutional law'.

This chapter investigates this definitional struggle as it unfolded during the World Summit on Information Society +10 (WSIS+10) review process in 2015, a focal point of conflict in the Internet governance global arena. Combining Hajer's methodology of discourse analysis with the narrative policy analysis approach, a content analysis of documents presented in five different stages of the WSIS+10 review process is applied in order to shed light on the process by which actors coalesce around common definitions and narratives, and eventually produce discursive orders.

THEORETICAL AND METHODOLOGICAL FRAMEWORK

Traditional policy studies based on a rational choice paradigm have usually been dealing with policy problems as if their meanings were clear and self-evident. This attitude has led social scientists to reify the dominant social groups' point of view or their own implicit assumptions and value judgments about the phenomena they were researching. Since the linguistic and argumentative turn, many scholars have underlined the role of ideas, conceptualizations, and interpretations within the policy process (Fischer 1993, 2003).

From this perspective, policy problems are conceived of as social constructs produced by the interaction among social actors and lying on moral or ideological stances. These social constructs shape the perceived interests of the subjects, and at the same time, they are also a function of particular material and social conditions (Edelman 1988, 9). Accordingly, the policy process could be described as a struggle for the definitions of a problem and of its boundaries; for the establishment of categories used to describe the policy problem, its causes, and effects; for the setting of criteria used for their classification and assessment; for the adoption of symbols, meanings and ideals evoked to guide and legitimize particular actions (Maynard-Moody and Kelly 1993, Stone 1988). Several authors have studied these definitional struggles as social discursive interactions that shape policymaking. In particular, Hajer (1993, 1995) has adapted the Foucauldian concept of 'order of discourse' to the policy studies field. According to him, a policy discourse refers to the tools and meanings through which actors produce and reproduce a policy problem as a social construct. Moreover, the discursive interaction also shapes the identities of actors, providing them with 'subject-positions' that define their social and power relationships. Concepts such as discourse structuration and discourse institutionalization play a central role in this perspective. Discourse structuration occurs 'when a discourse starts to dominate the way a society conceptualizes the world' (Hajer 1993, 46). Moreover, since policy discourses are deeply rooted in discursive practices and in historically determined institutional contexts, we can talk of 'discourse institutionalization' when a discourse 'solidifies into an institution, sometimes as organizational practices, sometimes as traditional ways of reasoning' (ibid.). Therefore, a policy domain could be considered as a battlefield where different policy discourses compete to conquer hegemony over the conceptualization and the institutionalization of collective action. More precisely, in each policy domain we can detect processes of cooperation and conflict between different discourse coalitions, which deeply affect all the stages of policymaking, from problem definition to decision-making and implementation. Following Hajer, a discourse coalition is defined as an ensemble of actors sharing a set of storylines, that are 'generative sort[s] of narrative that allow actors to draw upon various discursive categories to give meaning to specific physical or social phenomena' (Hajer 1995, 56) and 'provide actors with a set of symbolic references that suggest a common understanding' of the policy problem (ibid., 62). Storylines, for Hajer, are the discursive glue that keeps a discourse coalition together.

The concept of discourse coalition gave a significant contribution to policy studies allowing policy analysts to examine how different groups of actors confront each other to define and redefine collective problems, their solutions, and their own positions in the discourse that structures the policy domain.

Differently from other approaches, such as the Advocacy Coalition Framework, which identifies coalitions on the basis of a common policy belief system, the discursive approach grounded on storylines helps to explain how actors with different backgrounds, interests, social and cognitive commitment can be part of the same coalition. Indeed, storylines 'interpret events and courses of action in concrete social contexts . . . and symbolically reflect the concerns of core beliefs rather than the beliefs themselves' (Fischer 2003, 102). Moreover, unlike policy belief systems, storylines are relatively flexible structures. They open up space for policy change, and for cooperation and compromise. Furthermore, while policy belief systems require a 'non-trivial degree of coordinated activity' (Sabatier 1993, 25) among members of an advocacy coalition, storylines allow consideration for a variety of actors that do not necessarily recognize each other as a part of the same discourse coalition. Put differently, actors belong to a discourse coalition if they resort to common discursive elements and contribute to reproduce the same policy discourse.

In this study we integrate Hajer's discursive approach with the work of other scholars that have analytically described the structural elements of policy narratives. Jones, Shanahan, and McBeth (2014) have transposed some elements of the structuralist narrative approach from the field of literary theory to the field of policy studies in their Narrative Policy Framework (NPF). They assume that policy statements often present a narrative structure in order to persuade other actors, public opinion, and decision makers. Summarizing previous contributions (in particular Stone 2002, Ney 2006, Verweij and Thompson 2006), they claim that a policy narrative is characterized by the presence of:

i. a setting or context (taken-for-granted facts and information about the problem and different kind of unmovable constraint to the policy action)
ii. a plot that introduces a temporal element (beginning, middle, end)
iii. characters such as fixers of the problem (heroes), causers of the problem (villains), and victims (those harmed by the problem)
iv. a moral of the story, where a policy solution is normally offered.

In particular, characters have shown to play a very important role in this approach because of their strategic and persuasive function. Following Stone (2002, 2012) the NPF approach assumes that 'villains' and 'victims' are often employed for constructing different kinds of 'causal mechanisms', that are 'types of strategies used in structuring policy narratives to describe a relationship between a policy problem and its asserted cause' (Shanahan, Adams, Jones, and McBeth 2014, 70). Moreover, they underline actors' use of characters to strategically increase polarization and intensity of the conflict aiming at mobilizing support (Shanahan, McBeth, and Jones 2011).

Zittoun (2014) recalls very similar narrative elements in his conception of policy statements, which are distinguished in a policy problem and a policy solution conceived of as two, at least analytically, distinct constructs. Drawing on previous studies (Cohen, March, and Olsen 1972; Kingdon 1995), Zittoun underlines how often actors' preferences for a certain policy solution precede the definition of the policy problem itself. Accordingly, policy problems and policy solutions need to be 'coupled' together, for example, linking the supposed cause of a problem with the supposed consequences of a policy. Policy problems and policy solutions are assembled together during discursive interactions in which actors give rise to what Lindblom (1990) called 'mutual adjustments'—that is, a system of reciprocal manipulation through persuasion and anticipation of criticisms. Differently from Zittoun, we do not hold that actors belong to a discourse coalition only if they consciously adhere to a public, coded, policy statement. Rather, following Hajer, we consider that the sharing of some discursive elements is a sufficient condition to be considered as a member of a given discourse coalition. Nevertheless we retain that Zittoun's analytical framework, as well as his consideration on 'assembling' and 'mutual adjustment' processes, are very useful in order to understand how different discourse coalitions, and different actors within a discourse coalition, can reach an agreement and a common understanding within decision-making processes such as international summits or conferences, multistakeholder fora, and so on.

In addition to these structural narrative elements, we consider a further element of storylines that is able to grasp also their substantive content. As we have described before, in every policy field, and particularly in emerging complex fields such as Internet governance, actors select and emphasize some aspects of the social phenomenon addressed by a policy initiative in order to promote particular definitions of policy problems and solutions. In our case, relevant definitions concern the Internet itself, and Internet governance as a policy domain.

Summarizing, in this chapter we will conceive of storylines as analytically distinguished in:

a. a set of basic definitions, that in the Internet governance case consists of:

   i. conceptions about the nature of the Internet
   ii. definitions of the boundaries of the Internet governance as a policy domain

b. a cluster of structural elements defining the policy problem, composed of:

  i. the problematic, that is the reason why a particular situation is conceived of as a problem.

  ii. the identification of a social group of victims damaged by the problem

  iii. the designation of the causes of the problem

  iv. the characterization of the 'guilty party', that is, the group which is to be blamed for producing the problem

c. a cluster of structural elements defining the policy solution, such as:

  i. the action to take to solve the problem, that is the concrete policy instrument that actors desire to implement

  ii. the supposed consequences of those actions

  iii. the public of beneficiaries

  iv. a legitimized authority responsible for the solution (corresponding to the 'hero' in narrative policy approaches[1]).

Using this conceptual framework, we analyzed seventy-eight written contributions to the WSIS+10 review process from actors belonging to five stakeholder groups (governments, private sector, civil society, technical community, and international organizations). We chose the WSIS+10 review process because, as mentioned above, the institutional dimension plays a very important role within discourse analysis. Discursive struggles do not arise in a social vacuum but take place within an institutional context (Hajer 1995, 60). Discursive orders are produced and reproduced through institutional practices, and one discourse could be considered hegemonic in a given domain when it is translated into institutional arrangements. We opted for the World Summit on the Information Society, inasmuch as it is one of the most important arenas in the IG ecosystem. Moreover, we focused on WSIS+10, the overall review process about the implementation and outcomes of ten years of WSIS activities, because it constitutes the most profitable opportunity in recent years to observe what Hajer calls 'interpellation', that is one of those moments when routine proceedings are interrupted, a dominant order can be contested, and competing discourses can arise.

The coding procedure was performed as follows:

  i. authors have jointly hand-coded definitions and structural elements of storylines described in this section within the selected documents

  ii. four main different sets of definitions and storylines have been identified within the documents and assembled into coherent discourses through a hermeneutic process

iii. finally, each actor has been assigned to an ideal-typical discourse coalition on the basis of distinctive discursive elements represented in its documents

Lastly, we analyzed discursive interactions and negotiations that led to the WSIS+10 outcome to identify definitions and narratives that were successful in gaining place in the final resolution (A/RES/70/125).

THE WORLD SUMMIT ON THE INFORMATION SOCIETY AND MULTISTAKEHOLDERISM AS AN INSTITUTIONALIZED DISCURSIVE ORDER

Before we proceed with the empirical analysis of WSIS+10 documents, it is necessary to briefly address the origin of the overall process and the main controversies it faced during one decade. Subsequently, we identify the main coalitions of actors that struggled inside this arena, and the discursive order that finally arose from discursive interaction.

The World Summit in Information Society is a process launched in 1998 by the International Telecommunication Union (ITU), a UN agency, which led to two international conferences, the first one in Geneva in 2003 and the second one in Tunis in 2005. Almost all scholars acknowledge it as a pivotal point in the history of Internet governance, which steered toward a new governance model, and we add to a new discursive order: multistakeholderism (Hofmann 2007, Mueller 2010). Even if the summit concerned the broader concept of the information society, Internet governance issues soon became one of the most contested and controversial topics in the preparatory phase. Two points raised the most of discussions and disagreements. The first one concerned the most proper governance arrangement for the Internet: Who should define policy for the whole Internet? What is the role of nation-states, or more in general, of public authorities, in this process? The second contested point related to the boundaries of Internet governance as a policy domain: should it include only technical matters, or should it be extended to the political and social implications of ICT developments? These two points have marked out Internet governance debate since the early stages and seem to be strictly interconnected to each other.

As Mueller noted, controversies in Internet governance have been characterized by the opposition between those who believe that Internet rules should arise from spontaneous cooperation and free adoption, following a 'networking' governance model, and those who believe that 'governance emerges from adherence to rules enforced by an authority' (Mueller 2010,

257). This opposition first came forward during the nineties, in the so-called Domain Name System (DNS) war (Kurbalija 2014), a struggle over the control of domain names and addresses. In that case, a coalition including the US government and ICT corporations defeated the International Tele-communications Union (ITU) and a group of national governments that supported its call for a traditional intergovernmental governance arrangement based on international treaties. At the end, DNS was entrusted to the Internet Corporation for Assigned Names and Numbers (ICANN), a private nonprofit organization under the oversight of the US government. This solution meant the consecration of the so-called 'self-governance regime' (Hofmann 2007) and the affirmation of a hegemonic 'neoliberal discourse' (Chenou 2014). Neoliberalists rely on a narrative representing a bottom-up, market-driven coordination, with a leading role of the private sector, as the best possible mode of governance for the Internet. It is worth noting that one of the most relevant arguments of the US-led coalition was based on the assumption of the purely technical nature of the Internet. Indeed, self-governance supporters argued that whereas the Internet is a fundamentally technical matter—entailing the management of a transnational and decentralized architecture—state intervention should be avoided because governments' slow-moving, bureau-cratic, centralized approach would threaten the functioning of the Internet (Drake 2004).

During the preparatory phase of WSIS, a coalition of governments, gathering the BRIC and developing countries, questioned both these points. They rejected the usual narrow conception of Internet governance as limited to domain names, addresses, and protocols, and demanded the inclusion of issues such as digital divide, Internet access, interconnection costs, and cultural diversity. Moreover, they criticized the governance structure presented by ICANN, considering the overseeing role of the United States as unacceptable. These countries, usually defined as 'sovereigntist' or 'realists' (Mueller, 2010) demanded effective, legally binding decision-making processes at the global level, the enforcement of the Westphalian principle that states are equal in international law, and a substantial leadership of national governments and intergovernmental organizations in developing and ruling the Internet at the domestic level. Finally, a third coalition composed of civil society and international organizations raised its voice over human rights violations and called for rules and principles able to limit power abuses both by states and corporations, and to assure a wider participation in IG decision-making processes.

Tensions among actors, especially between sovereigntists and neoliberals from governmental delegations, were intense and endangered the continuation of the summit. In the end, participants found an agreement on the multistakeholder model of governance. Multistakeholderism first arose within the context of environmental policy arenas to indicate a 'participatory turn' in the global governance (Bäckstrand 2006), leading to new forms of collaboration between governmental and nongovernmental actors. In short, 'multistakeholderism entails two or more classes of actors engaged in a common governance enterprise concerning issues they regard as public in nature' (Raymond and DeNardis 2015, 3). As observed by many scholars, within the WSIS process, multistakeholder governance constituted a compromise between the opposing options for private or public regulation of the Internet (Mansell 2007, Raymond and DeNardis 2015, Hofmann 2016).

The main outcomes of the WSIS process—the Geneva Declaration of Principles and the Geneva Plan of Actions, the Final Report of the Working Group on Internet Governance, the Tunis Agenda for Information Society, and the Tunis Commitment—clearly show how an emerging discourse on

Table 4.1. Multistakeholder Discursive Order

Basic Definitions	
Conception of the Internet	Global facility available to the public
Boundaries of Internet Governance	Technical and public issues, especially access and digital divide

Policy Problem		Policy Solution	
Problematic	Improve participation Improve ICT development Reduce digital divide	**Action to take**	Multilateral, transparent and democratic governance, with the full involvement of governments, the private sector, civil society and international organizations
Victims	Developing countries, people excluded by the Internet	**Authority**	All stakeholders in their role and responsibilities
Causes		**Consequences**	Better development of the Internet, and better exploitation of the Internet for developmental purposes; inclusion of marginalized groups
Guilty		**Beneficiaries**	Developing countries, people excluded by the Internet

Internet governance was developed through 'assembling' and 'mutual adjustment' processes.

In fact, as shown in Table 4.1, these documents define the Internet as a 'global facility available to the public' that is a way to keep together both the ideas of the Internet as a commercial service and as a public good. Moreover, they extend the boundaries of IG, acknowledging that it 'encompasses both technical and public policy issues'. Also, the documents refer to the necessity to improve participation and to reduce digital divide as the main problematic of Internet governance. Developing countries and marginalized groups are identified as the public of victims in the information society The documents do not clearly identify the causes of the problem or the 'party of guilt', and this is a clear sign of the will to avoid politicization and further diplomatic disputes.

On the side of policy solutions, all these documents propose to develop forms of coordination among stakeholders. In so doing, they recognize governments, private sector, civil society, and international organizations as equally legitimate participants in the management of the Internet, each one in its respective 'role and responsibilities'. WSIS outcomes assign a special role to governments in dealing with international public policy issues (as sovereigntists had claimed), but not in the 'day-to-day technical and operational matters, that do not impact on international public policy issues' (as requested by neoliberals).

It is worth noting that, in WSIS' multistakeholderism, participation is conceptualized as a principle for a purely informative model of policymaking, where there is no place for any form of binding decision. Par. 72 of the Tunis Agenda that called on the UN secretary general to establish an ad hoc institution—the Internet Governance Forum (IGF)—is emblematic of this point. The Agenda in fact depicted IGF as a 'forum for multi-stakeholder *policy dialogue*', whose mandate is 'to *identify* and *discuss* public policy issues, *facilitate* discourse and the exchange of information and best practices, advise stakeholders and enhance their engagement' (emphasis in italics is ours).

It is a very 'different approach compared to the more outcome-oriented notions [of multistakeholderism] common in other policy fields' (Hofmann 2016, 35). Thus, although the outcomes of the WSIS process adopted most of the claims made by sovereigntists and civil society groups; the concrete governance arrangements they designed did not put into question the neoliberal status quo. Moreover, the agreement between the two sides was favored by a high degree of ambiguity. As many scholars have underlined, ambiguity 'blur or hide problematic implication of decisions' and 'allow participants to read themselves in collective programmes and actions' (Fischer 2003, 63, see also Stone 1998, 2002; Zittoun 2014, Cobb and Elder 1983). In this case, crucial elements such as the roles and responsibilities of different stakeholders were not well defined, and a pivotal expression such as 'day-to-day technical and operational matters that do not impact on international public policy issues' left ample room for interpretation.

Despite these limitations, or precisely because of them, the shift toward a multistakeholder model of governance received a broad consensus, it raised enthusiastic expectations and it rapidly spread in other important IG institutions, including ICANN itself (Hofmann 2016). Multistakeholderism became the dominant discursive order in WSIS environments (and more generally in the broader Internet governance ecosystem) fulfilling the criteria of structuration and institutionalization of a discourse within a policy domain as discussed above. Indeed, almost all actors accepted the categories of the multistakeholderist discourse to refer and give meaning to the 'Internet governance' phenomena and those categories solidified in WSIS organizational practices.

Nevertheless, the wide consensus surrounding multistakeholderism does not mean that Internet governance has turned into an appeased field.

Rather, it should be considered true what Hajer observed about environmentalism: when all parts declare to be 'Green' it means that the conflict has become fully discursive and based on interpretation and definitional struggles. Similarly, multistakeholderism is supposed to work as a discursive umbrella under which different discursive coalitions fight each other to impose their own definition of the basic terms of the matter. The ambiguity of the WSIS final statements, on one hand, facilitated compromise, but on the other, it kept several points of contention still open (Palladino and Santaniello 2021).

In the next paragraph, we are going to observe how different discursive coalitions exploited the possibility offered by the WSIS+10 review process—which started ten years after the Tunis WSIS—to impose their own narratives in order to redefine the meaning of some pivotal elements of the multistakeholder discursive order.

WSIS+10: RECONSTRUCTING DISCOURSE COALITIONS

Paragraph 111 of the Tunis Agenda asked the UN General Assembly to undertake the overall review of the implementation of the WSIS outcomes in 2015. In response, the General Assembly with resolution n.68/302 decided that the overall review would be preceded by an intergovernmental process taking into account inputs from all relevant stakeholders of the World Summit on the Information Society. Its main output was a draft approved by the General Assembly as a resolution on 16 December 2015 (A/RES/70/125).

Table 4.2.   Discourse Coalitions in WSIS+10

Neoliberals	United States(gov), Australia(gov), Canada(gov), European Union(gov), Japan(gov), Netherlands(gov), Poland (gov), ICC(ps), AfICTA(ps), IGC(ps), JBF(ps), JISA(ps), Telefonica (ps), Public Knowledge(cs), LIRNEasia(cs), WIPO(io), ICANN(ta), IEEE(ta), ISOC(ta), JPRS(ta), RIRS(ta)
Sovereigntists	China(gov), G77(gov), Russian Federation(gov), Egypt(gov), India(gov), Saudi Arabia(gov), Brazil(gov), Indonesia(gov), Iran(gov), Cuba (gov), ACNU(cs),
Constitutionalists	Access(cs), APC(cs), APIG(cs), ASEBE(cs), CDT(cs), CIS(cs), GPD(cs), IDP(cs), IRCP(cs), Justnet(cs), UNESCO (io), 4s(ta), CST-FVG(ta)
Developmentists	Argentina(gov), ASIS(gov), Asabe Sheu Yar'Adua Foundation (cs), Republic of Korea(gov), Paraguay(gov), Mexico(gov), Srilanka(gov), Azerbaijan(gov), Columbia(gov), Kenya(gov), ASDF(cs), FESF(cs), Huan Health(cs), IFLA(cs), OCCAM(cs), SEMANTIS (cs), ELAC(io), FAO(io), CDEUNDP (io), UNWOMEN(io), INDIANINSTITUTE(ta)
Not Assigned	25th Century Tecnology Limited (ps), Ark Earth Foundation (ps), Ec-Medici Framework (cs), ESCWA(nd), FANCV (nd), GLOBALFUND (nd), JPNIC (ta), OCCAM (cs), Pasa (cs), Switzerland (gov), TKEB (ps), Turkey(gov), Worldview Mission (nd)

Stakeholder groups: Government (gov), Private Sector (ps), Civil Society (cs), International Organization (io), Technical-Academic Community (ta), Not Declared (nd)

The WSIS+10 review process was another opportunity for actors to dispute Internet governance problems, to try to persuade each other, and to reach their own favorite policy outcomes. In this paragraph, we are going to illustrate the discourse coalitions we traced back following the methodology described in section 2. In particular, we identified four discourse coalitions in the WSIS+10 review process (Table 4.2): Neoliberal, Sovereigntist, Constitutionalist, Developmentist.

Only fourteen submissions out of seventy-eight were not assigned to any discourse coalition, mainly due to either their high level of ambiguity or their scarce salience for IG. It should be noted that some relevant actors in the WSIS process, such as Switzerland and the ITU, are not assigned to any coalition as their storylines fully adhere to the multistakeholder discursive order presented in Table 4.1, and they avoid using definitions and narratives which are distinctive of one or another particular coalition.

Moreover, we reconstruct the way actors reached an agreement on the outcome in order to investigate mutual adjustments among actors and the elements upon which the agreement was built.[2]

The Neoliberal Coalition

The neoliberal discourse coalition (Table 4.3) includes the national governments of the United States and developed countries, private sector associations and technical communities, and some civil society associations. Its conception of the Internet interlaces technical and economic aspects, as testified by the contribution of the Institute of Electrical and Electronic Engineers (IEEE), according to which the Internet rests on 'technical standards developed in an open, market-driven paradigm . . . these standards are voluntarily adopted and their success is determined in the marketplace'. Moreover, governments of the neoliberal coalition often recall the open, global, and interoperable nature of the Internet, where the reference to these elements has the function to warn other actors against any attempt to create alternative DNS and root zones, which would lead to a detrimental fragmentation or balkanization of the Internet. They also stress the importance of the private sector in the development of the Internet and underline the role of the market in fostering technological innovation. It is worth noting that the neoliberal coalition identifies as its own public of victims marginalized groups, minorities, and developing countries excluded from the social and economic benefits of the Internet. That is the same 'public of victim' depicted by the BRIC and other developing countries in their criticism of the US unilateralism during the preparatory phase of the first WSIS in 2003.

Table 4.3. Neoliberal Discourse Coalition

Basic Definitions	
Conception of the Internet	Global facility with technical standards developed in an open, market-driven paradigm
Boundaries of Internet Governance	Both technical and public policy issues

Policy Problem		*Policy Solution*	
Problematic	Development of the Internet	**Action to take**	Liberalization, light regulatory frameworks, low taxation, intellectual property protection
Victims	People excluded by Internet access	**Authority**	Public Private Partnership
Causes	Regulatory bottlenecks, protectionist policies	**Consequences**	Spur innovation and access
Guilty	States, Intergovernmental organizations	**Beneficiaries**	People excluded by Internet access

Nevertheless, at the same time, neoliberals define governments as the 'guilty party' that restrains Internet penetration by establishing legal barriers for private investments and protectionist policies. Indeed, we can read in the US contribution: 'One impediment to the build out and expansion of ICT networks and technologies is the existence of national and local policies that stifle or outright prohibit innovation and investment, including policies that forfeit progress to protect outdated, twentieth-century technologies, increase the power of governments, and/or restrict the free flow of information'. The replacement of the problematic of Internet governance formulated by their opponents (access, digital divide, and development issues) with its supposed causes (protectionist policies and regulatory bottlenecks) allows this coalition to fully deploy a typical neoliberal agenda at the core of the WSIS+10 review process. Thus, the proposed solution to face these problems is to protect the leading role of nongovernmental actors in existing governance structures and promote market-friendly policies. Moreover, such arguments are employed to redefine and limit the role and responsibilities of governments to the creation of an 'enabling environment' for ICT development, one of the tasks that the Geneva Action Plan assigned to them. From the neoliberal point of view, such an enabling environment consists in the public support for private investments and market competition. It includes a 'light touch', 'predictable' and 'non-discriminatory' regulatory framework, low taxation, intellectual property protection, and further privatizations. From the neoliberal perspective, such policies should let 'the private sector to continue to drive innovation to bring the Internet promise to more communities, and to help communities that are already online to take the better advantage of that access' (International Chamber of Commerce). Neoliberals firmly reject any form of internationalization and intergovernmental arrangements of Internet governance, recalling classic commonplaces on governmental inefficiency: 'the creation of new intergovernmental authorities or institutions to regulate the Internet would also result in rigid procedures, bureaucracy, political stalemate, and wasted time, which could stifle the innovation needed to develop solutions' (US). The coalition uses these arguments also to set some limits on the WSIS process itself. For example, the US government asserts that WSIS 'should avoid seeking to resolve ICT policy issues which are being addressed in other, more appropriate fora or re-opening issues concluded elsewhere'. That is a way to strengthen the status quo and the leading position of ICANN in the management of Internet critical resources. Finally, neoliberals present the Internet development as a successful story based on private entrepreneurship and self-regulation that should be continued in the usual way.

The Sovereigntist Coalition

The sovereigntist coalition (Table 4.4) includes BRIC countries (Russian Federation, China, India, Brazil), the G77[3], and some other developing countries. The coalition does not propose any specific conception of the Internet. Rather, they strategically refer to the broader definition of 'people-centred, inclusive and development-oriented Information Society'. The emphasis on people, inclusion and development is useful to recall government responsibilities, and to support the aim to bring back the Internet in the common field of other telecommunication systems, such as telephone, radio, and satellites, which are usually under intergovernmental regulation. Indeed, in their view,

Table 4.4. Sovereigntist Discourse Coalition

Basic Definitions	
Conception of the Internet	People-centered, inclusive and development-oriented information society; National strategic asset
Boundaries of Internet Governance	Technical and public issues, especially security

Policy Problem		Policy Solution	
Problematic	Inequalities, Exclusion, Impossibility for States to fulfill their responsibility in existing governance arrangements	**Action to take**	Intergovernmental management
Victims	Developing countries, population excluded by Internet access and benefits	**Authority**	Governments
Causes	Exclusion of governments by the technical management of the Internet	**Consequences**	Better development of the Internet, and better exploitation of the Internet for development purposes; inclusion of marginalized groups; strengthening of citizens and users' security
Guilty	Private sector management, US unilateralism, terrorists, and security breachers	**Beneficiaries**	Developing countries, people excluded by the Internet

the Internet, as other telecommunication systems, is above all a national strategic asset that states must control to guarantee the wellness of their respective populations. For the same reason, this coalition particularly stresses
security issues, much more than other coalitions. Sovereignists claim that
states have the mandatory duty to safeguard the security of networks, of their
citizens, and of the public order. The IG problematic is based on inequalities and exclusion, and the public of victims consists of all marginalized and
disadvantaged groups excluded from the benefits of the Internet. However,
they highlight that the 'vast majority of people that remain as yet uncon-
nected, including people with disabilities, to the Internet reside in developing countries' (G77 and China). Thus, they tend to present the developing
countries themselves as the main victims. The coalition identifies the cause
of the problem in the exclusion or marginalization of national governments
of developing countries in the existing governance mechanisms. Sovereigntists complain that recommendations enshrined in the Geneva Plan of Action
and the Tunis Agenda—which called for full involvement of governments in
Internet governance—had not been implemented yet, making it difficult to
address Internet-related problems. The US unilateralism in Internet governance appears as the main 'guilty party'. Indeed, several coalitions' members
recall that article 46 of the Geneva Declaration of Principles calls states to
'the avoidance of, and refrain from, any unilateral measure not in accordance
with international law and the Charter of the United Nations that impedes the
full achievement of economic and social development by the population of
the affected countries, and that hinders the well-being of their population'.
However, the US government is never mentioned directly. Regarding the
policy solutions proposed, we could observe two similar but different positions in this coalition. Both call for:

i. effective acknowledgment of States' sovereignty on Internet public
   policy and management
ii. full respect of the principles of international law enshrined in the UN
   Charter, particularly the political independence, territorial integrity and
   sovereign equality of states, and the principle of non-interference in internal affairs of other states
iii. a leading role of the United Nation or its agencies (especially ITU) in the
   internationalization of Internet governance.

These two positions differ in relation to the conception of the role of states
in multistakeholder governance. China and the G77 demand the strengthening of governments in multistakeholder institutions (above all ICANN),
in order to ensure 'equal participation' of all stakeholders, 'reasonable

allocation of Internet resources, and joint management of key Internet infra-structure'. According to the G77, 'multistakeholder model should not be lopsided, and any tendency to place sole emphasis on the role of businesses and nongovernmental organizations while marginalizing governments should be avoided'. These governments are asking for a 'participation on an equal footing' with other stakeholders in the technical management of the Internet critical resources, as well as a leading role of governments and intergovern-mental decision-making processes on some other public policy issues, espe-cially security-related ones. Instead, the Russian Federation calls for a leading role of governments even in the management of Internet critical resources. Indeed, the Russian contribution refers to a 'process of gradual transfer of control over the domain names and IP addresses allocation functions towards the international control under the auspices of the United Nations (or its specialized organization)'. Moreover, Russia pushes for a full localization of Internet resources, even at cost of fragmentation, as in the case of data centers and hosting servers: 'we call upon States to implement the storage of personal data of their citizens inside the territory of their own countries, to place domestic servers serving national segments of the Internet and to develop other elements of the critical Internet infrastructure'.

The Constitutionalist Coalition

The constitutionalist coalition (Table 4.5) includes several civil society associations, UNESCO, and the academic community. It should be noted the absence of any governmental or private sector actors. The constitutionalist conception of the Internet recalls both the definition of information society as 'people-centered, development-oriented inclusive' and the definition of the Internet as a 'global resource which should be managed in the public inter-est'. Some members of the coalition propose a little variation: 'Internet is a global public good, which must be managed in the interests of all the world's peoples' (JustNet, Association for Progressive Communication). Even if the sovereigntist coalition adopted the same definition of information society and some sovereigntist governments referred to the Internet as a public resource, it should be noted that for constitutionalists these words have a different meaning. Sovereignists refer to people, public resources, and public inter-est to legitimize national and intergovernmental public authorities. Instead, constitutionalists, through these words, refer to net-citizenry, net-citizens' rights, and true democratic processes. Indeed, constitutionalists identify the main problem of Internet governance in 'a lack of democracy; an absence of legitimacy, accountability and transparency; excessive corporate influence and regulatory capture; and too few opportunities for effective participation

Table 4.5.   Constitutional Discourse Coalition Storyline

Basic Definitions	
Conception of the Internet	Global public good, which must be managed in the interests of all the world's peoples
Boundaries of Internet Governance	Technical and public issues, especially fundamental rights

Policy Problem		Policy Solution	
Problematic	Net-Citizenship and power abuses	**Action to take**	Constitutionalize the internet; participatory policy-making to promote people's control of social technologies
Victims	People, citizens or net-citizens	**Authority**	International Human Rights Law; UN; Ad hoc independent authorities.
Causes	Lack of democracy and participation; absence of legitimacy, accountability and transparency; excessive corporate influence and regulatory capture	**Consequences**	Limits to power concentration and abuses, empowerment of citizens
Guilty	Private corporations and national governments	**Beneficiaries**	People, citizens or net-citizens

by people' (Association for Proper Internet Governance). According to them, such distortions in governance arrangements have led to an unacceptable concentration of power in the hands of corporations and governments that has amplified preexisting inequalities and allowed systemic human rights violations. For instance, the Internet Democracy Project denounces that the 'development of the information society has by default served first and foremost the interests of big corporations and governments, rather than those of users and people in general'. Constitutionalists point at 'commercial exploitation of personal data' and at an expansive use of intellectual property laws and policies that led 'to an unjust allocation of information goods', thus restraining innovation and development (Centre for Internet and Society). APC adds: 'states and businesses around the world are complicit in acts of communications surveillance that constitute violations of the right to privacy and other rights, including through legal frameworks, business practices,

and even the provision of public services'. In this storyline, thus, governments and private sectors clearly act as the 'guilty party' whereas the public
of victims consists of 'people', 'citizens' or 'net-citizens'. Nevertheless,
among the victims we can also find developing countries, marginalized and
disadvantaged groups just like in other coalitions, but in the constitutionalist
discourse coalition a special attention is paid to the situation of journalists,
bloggers, political oppositions, and rights defenders threatened by surveillance and censorship. To solve these problems, the members of this coalition
propose a series of provisions that, conceived as a whole, could represent an
attempt to constitutionalize the Internet. Indeed, their claims reflect the main
principles of the classic doctrine of constitutionalism, even if the concept is
never explicitly mentioned in their contributions: i) to limit power concentration ii) to safeguard people from power abuses iii) to empower citizens
(Sartori 2000). Most of this coalition's members use their submissions in
order to call for:

i. the implementation and enforcement of international human rights law
ii. the full respect of the rule of law
iii. inclusive, democratic, transparent and accountable Internet governance
processes and participatory policymaking to promote people's control of
social technologies.

Anyway, in most of the contributions by constitutionalists, such proposals
remained as unspecified as mere declarations of principles. Nevertheless,
some members advanced more detailed and applicable measures. The association Access Now suggested the inclusion of the net neutrality principle in
national legislations. The Association for Progressive Communication called
for the establishment of an independent judicial body responsible for decisions on limitations of the freedom of expression online. The Association
for Proper Internet Governance (APIG) and JustNet called for treaty-level
provisions, under the oversight of the UN and the ITU, that should bind states
to the respect of fundamental rights, especially in the field of cybersecurity.
JustNet, moreover, proposed to resort to international treaties also to reform
ICANN, strengthening its political accountability and adherence to internationally agreed norms and policies.

The Developmentist Coalition

This coalition (Table 4.6) consists of governments of developing and least
developed countries, civil society associations and international organizations
concerned with developmental issues and programs. This coalition appears

Table 4.6.   Developmentist Discourse Coalition

Basic Definitions	
Conception of the Internet	Driver for social and economic development
Boundaries of Internet Governance	Technical and public issues

	Policy Problem		Policy Solution
Problematic	ICT for developmental purposes; Reduction of technological gaps	**Action to take**	Aid, funding, technological transfer, capacity-building programs
Victims	Developing Countries	**Authority**	UN, International community
Causes	Lack of resources and capability	**Consequences**	Fill the gap with developed countries
Guilty		**Beneficiaries**	Developing countries

as the least sophisticated one within the WSIS process. Its discourse is not fully developed. The coalition members share a common understanding of the Internet as a 'leverage' or 'driver' for social and economic development. They focus on ICT development policies in addition to access and digital divide issues, but they do not address the question of Internet governance, neither to legitimize current arrangements nor to criticize them. When some members pay attention to Internet governance, they reveal a very narrow and uncritical view, just claiming for a wider inclusion of developing countries in current governance mechanisms but without calling into question governance structures, outcomes, or efficiency. Rather, developmentists certify their digital gap with the developed world and blame their insufficient resources and expertise as the causes. Consequently, they appeal to the international community, especially UN agencies, for support and cooperation, and particularly for funding, technological transfers, and capacity-building programs.

The WSIS+10 Outcome: Toward A New Discursive Order?

In this last paragraph, we analyze the discursive interactions and negotiations that led to the WSIS+10 outcome—the resolution A/RES/70/125. What kind of mutual adjustments among discourse coalitions took place? Which discourse coalition was the most successful in imposing its own points of view? Did this final statement bring forth a new discursive order as Geneva and Tunis outputs did ten years ago? Unfortunately, we cannot answer these questions by resorting to written documents that could give testimony of changes in actors' positions during the drafting process, since transcripts and

Table 4.7.   Outcome Storyline (United Nations Resolution 70/125)

Basic Definitions	
Conception of the Internet	Global Facility Driver for social and economic development
Boundaries of Internet Governance	Technical and public issues

Policy Problem		Policy Solution	
Problematic	Improve cooperation to face challenges	**Action to take**	Improve cooperation in respective roles and responsibilities
Victims	Developing countries, marginalized groups, users confidence and security	**Authority**	Private Sector (for technical management) States (for cybersecurity)
Causes		**Consequences**	Inclusive information society, building confidence and security
Guilty	Terrorists	**Beneficiaries**	Developing countries, marginalized groups, users' confidence and security

minutes of the last meetings—where the drafts have been negotiated—were not disclosed[4]. However, we can observe how draft content has changed until the approved resolution, to trace which contributions were embedded in the outcomes at each stage of the review process.

During the first steps of the drafting process, sovereigntist and constitutionalist claims seemed to meet a growing acceptance. The former obtained some acknowledgements about the idea of Internet governance as a 'multilateral' process and about the need to preserve the main principles of international law, such as political independence, territorial integrity, sovereign equality of states, and non-interference in domestic affairs of other states. The constitutionalist coalition gained a clear stance in favor of the protection of the net neutrality principle, and a call for an international convention on cybercrime that should bind states to the respect of human rights. All these points disappeared from the final text, and the outcome turned decisively toward neoliberal positions.

Moreover, if we consider the storyline of the final resolution (Table 4.7), we can observe how the Internet is defined as a 'global facility'. Compared with the definition of the Internet contained in the outcomes of WSIS

2003–2005, the absence of any reference to the 'public' testifies to the success of the neoliberal discourse coalition in the definition setting.

The public of victims includes once again developing countries, their population and marginalized groups excluded by the benefits of the information society, but also users threatened in their confidence and security. As ten years before, the statement avoids identifying a specific problematic for Internet governance, or causes and guilty parties, with the notable exception of 'terrorist' and 'cyber-criminal' that are explicitly blamed and condemned. Rather, the resolution prefers to talk about 'challenges' that require an improvement of cooperation among stakeholders. Put in these terms, the Internet governance problematic is functional to describe the current multistakeholder governance arrangements and the last ten years of WSIS activity as a story of success, which then must be confirmed and improved in order to achieve old and new goals. Indeed, the final statement repeatedly reaffirms the support for the principles of multistakeholderism, for the cooperation among stakeholders in their respective roles and responsibilities as established in the Tunis Agenda. The excerpt where governments were explicitly excluded by the day-to-day technical and operational matters is absent, but paragraph 55, just at the beginning of the section dedicated to Internet governance, states: 'We recognize that the existing arrangements have worked effectively to make the Internet the highly robust, dynamic and geographically diverse medium that it is today, with the private sector taking the lead in day-to-day operations and with innovation and value creation at the edges.' Furthermore, the emphasis on security concerns represents the major innovation of the WSIS+10 compared with previous WSIS outcomes, and also the field where national governments gained a prominent role. Par.50, for example, assigns to governments the leading role in cybersecurity. Par.53 calls upon 'Member States to intensify efforts to build robust domestic security in information and communications technologies and the use thereof, consistent with their international obligations and domestic law . . . in combating the criminal misuse of the technologies and preventing the use of technology, communications and resources for criminal or terrorist purposes'.

Despite appearances, the acknowledgment of an important role for governments in security matters does not constitute a concession to sovereigntists. Rather, it could be considered as a meeting point between the two most important discourse coalitions of the WSIS+10—the neoliberal and the sovereigntist one—whose leading governments have always been asserting their own sovereignty on security issues, with no substantial difference between the US and China or Russia. The other two alternative discourse coalitions—the constitutionalist and the developmentist ones—have instead obtained nothing more than some symbolic acknowledgements in the final outcome. Constitutionalists gained just a generic commitment toward the respect of international human rights law and the condemnation of surveillance practices, but without the provision of any implementing mechanisms. Furthermore any arrangements intended to make the governance system more democratic and accountable to people were never really considered during the drafting process. Developmentists obtained just a generic commitment to support ICT development in developing countries, but no concrete financial or fiscal mechanisms.

CONCLUSIONS

In this study we have addressed global Internet governance from the point of view of discursive struggles engaged by several coalitions around controversial issues. We have identified four discourse coalitions through an analysis of contributions submitted to the preparatory phase of the 2015 WSIS+10 review process. Our analysis testifies how actually several actors belonging to different stakeholder groups (governments, private sector, civil society, technical community, international organizations), even in absence of previous and structured forms of coordination, could be traced back to a limited number of discourse coalitions. Indeed, almost all actors shared some discursive elements that can be attributed to a specific discourse among the neoliberal, sovereigntist, constitutionalist, or developmentist ones. This means that Internet governance domain and controversies are structured by the tension between these discourses, and that actors reproduce these tensions in their discursive interactions. Moreover, although the Internet governance domain appears to be dominated by a widely accepted multistakeholder discursive order, we observed how the four discourse coalitions tenaciously struggled to redefine the meaning of some basic terms of the multistakeholderist discourse. Each coalition developed narrative strategies in order to legitimize its favored policy output and to define power positions by blaming their opponents and affirming its own authority or legitimacy. One of the most intense definitional struggles concerned the role and responsibility of different stakeholders. This point seems to make evident that the Internet multistakeholder governance model has not soothed conflicts regarding the role of public authorities, which has been one of the most debated controversies in Internet governance since its very beginning. Furthermore, our study leads to the conclusion that not all discourse coalitions have equal political weight within the multistakeholder order. The neoliberal coalition appears to be more proficient than others in advancing its own policy solutions within the process, even in a process led by intergovernmental organizations as the one we are analyzing. Of course, extra-discursive power resources play a relevant role in this scenario, but we can affirm that neoliberals exerted a discursive hegemony in the WSIS +10 review process. This hegemony can be detected in a discursive mechanism through which the neoliberal discourse coalition was capable of co-opting discursive elements from other coalitions, integrating them within its own narratives. Examples of this mechanism, as we have noticed in our analysis, are those of access and digital divide, that were used to bridge the neoliberal and the developmentist discourses in the attempt of providing neoliberal solutions to a developmental problem, i.e., removing protectionist policies and regulatory bottlenecks.

Concluding, this study highlights the crucial role of definitions and narratives in shaping global policy processes. The case of the WSIS+10 review process demonstrates that discourse coalitions tend to emerge in policy arenas where different kinds of actors are involved and that processes of meaning-making are valuable resources used by actors in their policy and diplomatic struggles.

NOTES

1. The structural elements composing policy problems and policy solutions are largely drawn by Zittoun, 2014.

2. All actors' submissions for the WSIS+10 preparatory phase are available at https://publicadministration.un.org/wsis10/nonpapersubmissions (Accessed June 11, 2021)

3. The Group of 77 (G77) at the United Nations is a coalition of non-aligned developing countries founded in 1964, to promote its members' collective economic interests and create an enhanced joint negotiating capacity in the United Nations.

4. The preparatory process roadmap can be found here: https://publicadministra tion.un.org/wsis10/roadmap/wsis10/roadmap

II

INTERNET GOVERNANCE AS A
SCIENCE DIPLOMACY AREA

Chapter Five

Science Diplomacy and Internet Governance

Opportunities and Pitfalls

Robin Mansell

This chapter examines opportunities and pitfalls associated with intersections between what has come to be designated as science or digital diplomacy and research in the internet governance field. Attention is given to the potential for conflict among proponents of differing views of appropriate institutionalized governance arrangements when they become active in diplomacy initiatives. This chapter also considers potential outcomes when members of the internet governance research community engage in diplomacy with a view to tackling socio-political challenges associated with the digital environment. It focuses particularly on differing views about the authoritative status of scientific evidence on the impacts of internet governance policies and practices and of the status of researchers in decision contexts that affect state, private sector, and civil society stakeholder interests in governing the digital ecology.

The digital environment, and particularly the internet, are increasingly sites of socio-technical controversy, and governance institutions are evolving rapidly. This is an area in which research is expected to provide insight in the face of very rapid technological innovation and market development of digital technologies and applications. The priority of digital platform operators is to expand markets and profits through global data collection and processing and these companies are operating in highly concentrated markets characterized by power asymmetries (Taplin 2017; Mansell and Steinmueller 2020). As gatekeepers or intermediaries, they have the capacity to block or filter digital information and to process customer data in ways that result in corporate or state surveillance. The financial strength of some these companies gives them a near monopoly and, therefore, substantial self-regulatory decision-making power. Their platform power is exercised through lobbying, which enables them to influence whether and how they are regulated by state actors or held to account by civil society actors (Couldry and Mejias 2019; Zuboff 2019).

The challenges facing those seeking to hold the digital technology companies to account are exacerbated by the framing of mainstream economic
science which suggests that formal regulation by states should be considered
only when there is scientifically verified evidence of narrowly defined market
failure, validated according to the norms of the prevailing science paradigm.
Such evidence is extremely difficult to produce and typically is contested
because of the complexities of the companies' operations and the digital
market structure (Daly 2016; Moore and Tambini 2018). The interests of
states, at least in the Western world, are both to secure access to data about
their citizens for security purposes and to achieve protections for citizens'
fundamental rights to access digital information, freedom of expression and
individual privacy (UNESCO 2021). These conflicting interests are making it
increasingly difficult to ensure that internet governance arrangements protect
citizens' interests.

Diplomacy and, specifically, science diplomacy, as a modality through
which states can pursue their interests in tackling socio-technical challenges
in the digital environment, is attracting increasing attention. In Europe, for
instance, European Commissioner Moedas, Commissioner for Research,
Science, and Innovation, has emphasized the role of science diplomacy in
responding to issues raised by climate change and the spread of infectious
diseases. The results of scientific research, including social science, are
treated in this context as a resource that can be deployed to exercise 'collective responsibility in a spirit of international solidarity . . . to solve common
and complex global challenges' (Moedas 2016, 9), although it may be argued
that science diplomacy does little more than signal a need for international
cooperation (Rüffin 2020).

The digital environment is a key site of socio-technical challenge that is
linked to economic competitiveness, security, and human rights in the European Union's strategic agenda (EC 2020) and similar linkages are present
in other countries and regions around the world (UN 2019). Researchers
working in the internet governance field are engaging with the diplomatic
community as numerous multistakeholder commissions and initiatives are
launched to tackle the risks and harms associated with the digital environment. In this chapter, the focus is on science, or what may be referred to as
'digital', diplomacy in relation to governing the consequences of the internet
and digital innovation. It does not address another sense in which the term
digital diplomacy is used, that is, in reference to the use of digital technologies by the diplomatic community to support their communicative strategies
(Bjola and Zaiotti 2020; Cerf 2020; Huang 2020).

The next section outlines contending theoretical perspectives on diplomacy
and its links to the prevailing paradigm of science, and to the less dominant

tradition of social constructivism in the social sciences. The third section provides illustrations of the digital environment as a site of socio-technical controversy. This is followed in the fourth section by a discussion of the emerging entanglements of science and governance and their association with alternative institutional models for governing through conferring authority, giving particular attention to multistakeholder forms of governance. This discussion is complemented in the fifth section by a focus on conditions that affect expectations about the influence of research produced by internet governance researchers when they engage in diplomacy. In the conclusion it is argued that the intersection of internet governance researchers with science or digital diplomacy does have the potential to influence governance outcomes in ways that may garner support for a digital environment in which citizens' fundamental rights are better respected, but it also acknowledges the pitfalls of such engagement.

DIPLOMACY AND SCIENCE AND TECHNOLOGY

The historical relationship between diplomacy and science (and technology) is ambivalent insofar as Nye's (1990) notion of soft power diplomacy, as distinct from hard or coercive power exercised by the military and in the economic sphere, includes the scientific, the cultural and the ideological. Nye (1990, 167) argued that diplomacy involves efforts to make state power 'seem legitimate in the eyes of others'. It is not neutral because it involves choices about values (Nye 2004). In the context of governing scientific and digital technological innovation, this implies that the engagement of researchers with diplomacy involves value choices and decisions about the authoritative status of research evidence as well as the respective authority of state, private sector, and civil society actors within the institutions developed for governance.

From the dominant perspective, the processes and practices of diplomacy are understood to occur in the context of a 'cosmopolitan democracy' where the emphasis is on the rule of law, representative decision-making, and formal accountability for decisions (Held 1995). Hierarchical arrangements for governance mean that certain stakeholders are accorded greater authority than others, and some actors will be excluded from decision making processes. Power relationships among state, corporate and civil society actors, including individuals with research expertise, are theorized in different ways, but power is typically conceived to work through pluralist relationships among actors who are assumed to engage with each other on an essentially level playing field. It is assumed that multiple viewpoints can be expressed, consensus achieved, and policy introduced that is regarded as the legitimate expression of the authority of the engaged actors (Lindblom 1990). This neo-pluralist notion of power provides a basis for assuming that a variety of actors—state, corporate, and civil society—is able to claim legitimacy in governance processes and to participate effectively in informing the outcomes. A pluralist view of power is predominant in the subfield of science diplomacy, defined broadly in the context of the United States, 'as scientific cooperation and engagement with the explicit intent of building positive relationships with foreign governments and societies' (Lord and Turekian 2009, np). In relation to the governance of the 'internet economy' in the American context, Zysman and Weber (2001, 15) found that associations between research in this area and diplomacy produce conflicts over 'fundamental values and basic choices about markets, community, and democracy'. In the European context, increasing attention is focusing on tensions between the protection of commercial or market values and public values in the digital innovation space (van Dijck, Poell, and de Waal 2018).

The dominant view of science diplomacy is consistent with a prevailing view of scientific research in which 'scientific values of rationality, transparency and universality are the same the world over' (The Royal Society 2010, vi). Science and social science are seen as providing 'a non-ideological environment for the participation and free exchange of ideas between people, regardless of cultural, national or religious backgrounds' (The Royal Society 2010, vi). This view grants authoritative status to all those who have the appropriate professional qualifications, and it allows scientists to insist on their independence from political, economic, or social influences within diplomacy processes. It is this construction of science that the United Kingdom's Royal Society employs when it characterizes 'science in diplomacy' as 'informing foreign policy objectives with scientific advice' which is valued for its contributions to an evidence base that can be used to support political decisions (The Royal Society 2010, v). 'Diplomacy for science' is seen as facilitating international science cooperation. 'Science for diplomacy' is then treated as using science cooperation to improve international relations between countries. In this context, science is depicted as value free, consistent with the prevailing conception of scientific inquiry. Scientific evidence is regarded as a resource that is 'complete' or final and 'the most objective thing known to man' as Einstein (1934/2009, 112) put it.

While the processes of scientific inquiry are assumed to be open to discovery and drawn from multiple sources, openness is not regarded as an unqualified good in this paradigm since 'there are legitimate boundaries of openness which must be maintained in order to protect commercial value, privacy, safety and security' (The Royal Society 2012, 9). This is deemed to be the optimal way to ensure that scientific results are 'assessable so that judgments can be made about their reliability and the competence of those who created them' (The Royal Society, 2012: 7). These institutional norms serve as a means of maintaining the privileged authoritative position of those researchers who adhere to the norms of this scientific paradigm (David, den Besten, and Schroeder 2010), with the result that there is reticent among scientists to disclose all that is known. The obligation is to disclose only enough information to enable others, such as diplomats, to interrogate research results (David and Steinmueller 2013). The result is a hierarchy that is maintained between those deemed qualified to offer authoritative views and those whose views are either discredited or downplayed. This hierarchical system also helps to preserve the notion that scientific evidence offered in the conduct of diplomacy is not associated with normative judgments.

In contrast, an alternative view holds that diplomacy always operates within the framework of a 'discursive democracy'. In this context, governance decision-making is understood to be embedded in social and political discourses among multiple stakeholders—states, companies, and civil society—with the aim of encouraging the articulation of diverse discourses and reaching decisions that resolve contradictory interests (Dingwerth 2014). In this view, however, diplomacy is understood to be associated with asymmetrical structural power which mediates governance through 'conceptual systems and cultures—forged and modified through institutional, organizational, and technological mediators' (Comor 1999, 119). This asymmetrical power can operate to suppress or negate the broader interests of civil society actors in designing governance institutions that aim to achieve social and economic justice (Fraser 2010). In this view, when diplomacy intersects with science and technological innovation, it is expected to privilege the power of the state or corporate actors and to produce outcomes that sustain the global capitalist economy (Chenou 2010; Strange 1998).

From a Foucauldian perspective on power relations, the engagement of science with diplomacy can therefore be expected to normalize existing asymmetries of power among interested stakeholders and to shape perceptions of the most appropriate governance arrangements to favor prevailing scientific norms that confer authoritative status on research evidence. In effect, researchers engaging in diplomacy will be disciplined to prefer governance arrangements and outcomes that replicate or reinforce societal inequalities (Comor and Bean 2012). Thus, when these power asymmetries are taken into account, neither the conduct of scientific research, nor the contributions of research expertise to diplomacy, can be assumed to be independent or value free. Notwithstanding their frequent claims to independence, researchers working in the prevailing scientific paradigm are acknowledged from this critical perspective to be providing normative interpretations of their research findings. These interpretations may align with state, corporate or civil society interests and this is likely to yield tension and conflict (Van Langenhove 2016b; Gieryn 1983).

The predominant depiction of science and its authoritative status as 'distinctively truthful, useful, objective or rational' is strongly contested in other traditions in the social sciences with implications for the engagement of internet governance researchers in diplomacy. In contrast to the prevailing view of science, in this tradition scientific knowledge is understood to emerge through a socially co-constructed process (Mackenzie and Wajcman 1999). In this sense, the outputs of science are always 'in the making' and science is 'an end to be pursued' (Einstein 1934/2009, 112). Science is not a process of producing universally valid 'truth' claims. Research evidence produces knowledge that is co-constructed by multiple actors through practices characterized by Callon and Rabeharisoa as 'research in the wild' (2003). In this framing, theories and the interpretations of empirical results are intricately interwoven with political, economic, cultural, and social values and goals and all knowledge construction is assumed to be infused with power asymmetries. It is ideological (Gieryn 1983) and, as Dewey insisted (Boydstone 1989), researchers cannot escape from the normative implications of their work. Thus, the scientific community is understood to operate as a change agent in society rather than as a neutral bystander. This is because researchers are understood to make choices about the salience of competing theoretical paradigms and acceptable interpretations of their research results, and they inevitably privilege certain values and outcomes (Cammaerts and Mansell 2020).

This social constructivist view of scientific practice developed in parallel to the prevailing norms of science in the post–World War II period. The constructivist view of knowledge generation encourages the deconstruction of the rationalist discourse of mainstream science, and it considers asymmetrical power relationships among actors engaged in diplomacy. In relation to research on socio-technical controversies such as those encountered in the digital realm, researchers can be expected to draw upon a variety of social science disciplines to frame their empirical work with a view to accumulating insights into the ways in which scientific and technological innovations are shaped by cultural, social, political, and economic factors (Bijker, Hughes, and Pinch 2012). Multiple actors from a wide range of stakeholder groups are expected to influence, for example, whether new digital technologies emerge from the laboratory, how they are brought to the market, and their socio-political and economic consequences. In addition, in privileging the co-construction of knowledge, researchers working in this tradition tend to favor inclusive, non-hierarchical research and governance practices. This means they often have a greater affinity with the interests of citizens in ensuring that policies designed to address socio-technical controversies are responsive to their needs and priorities, in contrast to the dominant scientific paradigm which asserts that research practices are value free.

Science or digital diplomacy is not a new phenomenon, but it is achieving a high profile in the face of global policy challenges and policy makers have a growing need to understand scientific evidence. Scientists and social scientists are increasingly being called upon to contribute to national, regional, or global policy in issue areas such as climate change, gene therapy, genetically modified organisms, heath, cybercrime and the uses of artificial intelligence and robotics. Debates in these areas often are politically charged and public controversy can lead to claims that science is not sufficiently insulated from politics (Jasanoff 1990; 2021). The authoritative status of research evidence and of researchers is increasingly being called into question and the 'science-policy relationship is sometimes difficult and occasionally dysfunctional' (Sutherland et al. 2012, 1).

This is so not only as a result of conflicts between communities of scientists who adhere to different paradigms of science, but also because some researchers are basing their claims to authority partly on their engagements with 'researchers in wild', that is, with a wide range of citizens (Callon and Rabeharisoa 2003). They also frequently are employing research practices that are underpinned by qualitative methods (or mixed methods) which are seen by those adhering to the predominant science model as less robust than quantitative experimental methods as means of validating research results. These non-dominant approaches are frequently regarded as producing evidence subordinate to that produced by formally accredited science (Stodden 2010). As Callon (2003) points out, 'faced with the exceptional' demands arising in the face of global socio-technical controversies, research-based insight may be generated by researchers who do not know each other well, for instance, through crowdsourcing, and who have a wide variety of methodologies for validating the authoritative status and interpretation of their knowledge and its implications for policy (Callon 2003).

Although scholars working in the 'subordinate' tradition argue that it is possible to 'to maintain a careful balance of scientific advice, stakeholder participation, public debate, and political discretion, which is crucial for handling the risks and benefits of modern technological cultures in a democratic way' (Bijker, Bal, and Hendriks 2009, 5), the social constructivist tradition in the social sciences is not immune to controversy. This is especially so with regard to the authoritative status of evidence and the normative commitments researchers bring to the diplomatic process since they often seek to improve governance institutions or to invent new ones. To succeed in implementing their ideas, they muster political support, and this results in entanglements with the dynamics of asymmetrical power relations among state, corporate and other civil society stakeholders and in conflicts over the values and goals that should receive priority in a given issue area such as internet governance (Franklin 2013a).

In summary, a pluralist view of the power which informs the dominant view of diplomacy is well-aligned with the prevailing paradigm of science, but its hierarchical norms and conventions are being contested in multiple areas. This is especially the case as the status of scientific inquiry is declining in the popular imagination in Western democracies (Bijker, Bal, and Hendriks 2009), notwithstanding claims to science-led policy making and the politicization of science in the face of the COVID-19 pandemic (Jasanoff et al. 2021). This context needs to be considered when the implications of engagements between researchers in the field of internet governance with science or digital diplomacy are examined.

THE DIGITAL ENVIRONMENT AS A SITE OF SOCIO-TECHNICAL CONTROVERSY

Conflicts around the governance of the digital environment are present globally, although the structures and processes for addressing them are specific to institutional arrangements in each region or country (Brousseau, Marzouki and Méadel 2012b; Brown and Marsden 2013; DeNardis 2014; Marsden 2017). In the global context, efforts to preserve an open, fair, and transparent internet that is consistent with the interests of citizens are being challenged in multiple ways and often around the concept of sovereignty (Mueller 2020). 'Internet freedom' is declining as governments seek access to data from social media and apps (Rogers and Luck 2017) and the market for abusive online content is growing (UN 2020). Internet fragmentation resulting from the absence of global agreement on internet governance issues is exacerbated when countries adopt policies and practices that are inconsistent with citizens' fundamental rights (UNESCO 2021). Fundamental rights to access digital information are curtailed, for example, when countries block access to social media (Marchant and Stremlau 2020; Shahbaz and Funk 2020).

Controversies over governance and its outcomes in the digital environment generate struggles over the authoritative status of research contributions provided by adherents to different scientific norms and lead to divergent views on institutionalizing governing authority as well as about the power of relevant stakeholders, including the state, companies, and civil society. Different epistemologies or paradigms of science influence what comes to be regarded by the diplomatic community as standard evidence and as 'normal' patterns of interaction and ways to resolve disputes (Nelson and Sampat 2001). In the dominant science community and among state and large corporate stakeholders, the preferred structure of governance arrangement can be described as constituted authority which involves formal and hierarchical norms and procedures for accumulating authoritative knowledge (Mansell 2013; Powell 2015). In the constructivist traditions of the social sciences, in contrast, the accepted norms and procedures for conferring authority on research evidence and researchers are less formal and non-hierarchical. They are more fluid and can be characterized as adaptive forms of authority. These distinctive approaches give rise to conflict which is especially visible when value conflicts are present and is accompanied by claims and counterclaims in governance settings about the authoritative status of research evidence (Cammaerts and Mansell 2020).

The instability of governance arrangements in the digital environment illustrates the conflictual nature of these socio-technical challenges and the politicization of scientific evidence. When, for example, President Trump signed a congressional joint resolution reversing the privacy rules that were to have applied to internet access service providers and had been put in place by the Federal Communications Commission (FCC) (2016), the rules had been informed by a wealth of research evidence. If the rules had come into effect, they would have required companies to obtain customers' permission prior to the use of data designated as 'sensitive', a move designed to provide improved protection of personal data. Some privacy protections remain in place (Drye 2017), but the multi-layered and interlocking ownership structures of the companies in the digital market mean that threats to privacy are now considerably greater than they would have been if the privacy rules had come into effect. The authoritative status of research evidence provided by proponents of the privacy rules was contested and, in this instance, overturned. This national set of deliberations also was influential in creating a discourse that found its way into global contexts and diplomacy initiatives (Hofmann 2020).

Within countries, the governance arrangements that historically have been used to protect the public interest in the face of the market power of digital network and service providers through regulatory intervention are increasingly open to challenge. States do, however, sometimes intervene to protect citizens' fundamental rights as in the case of efforts to preserve an open internet. For example, the FCC introduced network neutrality rules to secure an open internet, notwithstanding claims that this would jeopardize the competitiveness of internet access service providers. These rules were underpinned by a research evidence base supported by some researchers and criticized by others, all of whom made claims to the authoritative status of their evidence. After regulatory proceedings and court challenges, the companies providing access to the internet were required to give equitable treatment to internet traffic (FCC 2015). The instability of this form of governance was confirmed by the move by then FCC Chairman, Ajit Pai, to overturn the FCC's 2015 order (McKinnon 2017). This change favored the interests of some of the largest companies involved in content provision and data transmission, and of the government, because it allowed them to discriminate among different types of digital information for commercial or state security purposes. Similarly, any future change in position on the network neutrality issue will rely on research evidence and the research community will be called upon to make claims to the authoritative status of their competing theories, empirical methods and results. This was an instance of the contested entanglement of science with the process of governing in the digital environment and it too generated a discourse which can be, and sometimes has been, appropriated globally (Broeders and van den Berg 2020a). This entanglement of science and diplomacy is visible, for example, in the contemporary geopolitics of 5G technology deployments which are framed by a trade war between the United States and China and by claims and counterclaims about the impact of network virtualization on national security (Mansell and Plantin 2020; Tang 2020).

EMERGING ENTANGLEMENTS OF SCIENCE AND GOVERNANCE

The prevailing paradigm of science and state or corporate-led governance fosters structures and processes of deliberation on socio-technical controversies that typically favor constituted authority. Recently evolved governance institutions concerned with the internet and the broader digital environment, however, are more closely aligned with the norms of adaptive authority. Those who affiliate themselves with these norms are more likely to privilege values of openness and transparency than are those who affiliate with governance arrangements aligned with constituted authority (Mansell 2013; Mateos Garcia and Steinmueller 2008). In addition, the diplomacy institutions of constituted authority are starting to embrace some of the features of adaptive authority with implications for the outcomes of the interpenetration of research on internet governance and science diplomacy. In turn, this has implications for the authoritative status of the stakeholders (state, corporate and civil society) and, in the case of researchers, of their scientific evidence base. This intermingling of norms and practices and contests over what should be 'standard' patterns of interaction is occurring in multiple jurisdictions worldwide with consequences for whether citizens' interests are taken into account in resolving socio-technical controversies related to the digital world.

There are numerous modalities of governance in the internet domain, and the way in which authority is institutionalized differs considerably (Kleinwachter 2017), with some arrangements being closer to constituted authority and others to adaptive authority norms and practices. These differences are visible in multistakeholder forums where stakeholder participation ranges from the Internet Corporation for Assigned Names and Numbers (ICANN) model where governments are involved in advisory capacities, to the World Summit on the Information Society (WSIS) where nongovernmental stakeholders were involved, but only in consultation, and the Internet Governance Forum (IGF), where multiple stakeholders participate, but without formal decision-making authority. Internet governance issues (e.g., the digital economy, cybersecurity, and human rights) are addressed by organizations including the International Telecommunication Union, the World Intellectual Property Organization, and the United Nations Educational, Scientific and Cultural Organization, as well as the World Trade Organization and the World Economic Forum, all of which have characteristics of constituted authority despite their appropriation of the term, multistakeholder (Kleinwachter and Almeida 2015; Pohle 2016). A complex system of global governance is developing which can perpetuate, maintain, or challenge hierarchical power among the stakeholders in different contexts (UN 2019). The locus of authority and the perceived authoritative status of researchers depend on which norms and practices designed to confer authority are valued and privileged. All of these governance arrangements are increasingly drawing upon research evidence and, as such, they represent a diverse set of interconnected instances of science or digital diplomacy.

Internet governance research and practice can be defined narrowly in relation to internet resource management or, more broadly, in relation to the governance of content production and online interaction (Brousseau and Marzouki 2012). The adaptive authority bias is present in this dynamically evolving area which favors multistakeholder arrangements in which transparency and the equitable involvement of multiple stakeholders are valued. The result is that less attention may be given to the dominant scientific norms for establishing professional authority and there is greater openness and validation for the views of citizens than is typical in the constituted authority view of governance. Akin to commons-based forms of organization (Benkler and Nissenbaum 2006; Poteete, Janssen, and Ostrom 2010), the hallmark of the newer forms of multistakeholder internet governance is collaboration, dispersed initiative, fluidity, and rapid action. These models of internet governance are evolving largely from the bottom up and emphasize the principles of account-able forms of participation, ideally to achieve 'hybrid, bifurcated, plurilateral, multi-level, and complex modes' of governance (Bäckstrand 2006, 468). The challenges of achieving this ideal on a 'transscalar' and 'transcultural' global basis are considerable (Scholte 2014; 2002) and norms are being devised to provide means for managing conflicts about where authority is, or should be, located and what evidence counts as authoritative knowledge.

With the internet as a central component of the material and immaterial infrastructure for global mediated communication, the status of researchers and research evidence is crucial because of its influence over the democratic legitimacy of the institutions involved in the social ordering of a digitally enabled society. As Franklin (2013a, 183) puts it, 'like Rip Van Winkle, government regulators have discovered that things have changed and they no longer call the shots in terms of internet design, access, and use'. They do not have uncontested recourse to assertions about the authoritative status of scientific research evidence that they once called upon to sustain outcomes favoring state and corporate interests over those of citizens.

When researchers are called upon to participate in some of the governance arrangements that are characterized as multistakeholder, however, there is often only a weak commitment to reflexive or inclusive practice consistent with adaptive authority. For example, when multistakeholderism is referred to in a report prepared for the World Economic Forum, it is noted that 'there are strong divergences of views between governments and citizens about whether MSGs [multistakeholder groups] are near angels who can deliver everything or whether they are inherently dangerous' (Gleckman 2016, 94). Here, lobbying and advocacy to provide 'independent' knowledge to govern-ments is deemed appropriate for consultative processes, but the authority to take decisions about how to address global problems is restricted to stake-holders adhering to the constituted authority model.

A partial incorporation of the norms and values of adaptive authority is typical in other multistakeholder initiatives which are implicating research-ers in science or digital diplomacy. For instance, the G20 Hangzhou Sum-mit's Global Digital Economy Development and Cooperation Initiative was welcomed by some members of the internet governance community because of its commitment to 'a multistakeholder approach to Internet governance, which includes full and active participation by governments, private sector, civil society, the technical community, and international organizations, in their respective roles and responsibilities' (G20 2016, para 5). In this case, however, features of constituted authority governance that confer primary authoritative status on state actors were retained. The G20 countries called for open, transparent, inclusive, evidence-based policy making soliciting comments from public and private stakeholders, but consultation was positioned only as a prelude to state decision making. It should occur '*before* laws, regulations, policies and other instruments are deliberated, developed and implemented' (emphasis added) (G20 2016, para 14).

Similarly, the nongovernmental Global Commission on the Stability of Cyberspace established to tackle cyber warfare was envisaged as a global multistakeholder initiative. The aim was 'to convene key global stakeholders to develop *proposals* for norms and policy initiatives to improve the stability and security of cyberspace' (emphasis added) (GCSC 2017). Nye (2017) suggested that the Commission might have a better chance of reaching agreement on the governing norms for the use of cyber weapons than the United Nations Group of Governmental Experts because it does not grant states ultimate authority. He regarded this as potentially opening up a space for deliberation among a wider set of stakeholders, including the research community. However, in line with the predominant view of science diplomacy, the Commission received proposals from scientific experts and other stakeholders, while decision making authority rested with the state members for the production of the Commission's final report (GCSC 2019).

The state or corporate actors have principal authority as well in the model proposed by the Global Commission on Internet Governance. Its *One Internet* report called for an arrangement 'in which affected stakeholders who want to participate in decision making can, yet where no single interest can unilaterally capture control' (GCIG 2016, np), consistent with an inclusive pluralist notion of power. This suggests an aspiration towards an adaptive authority model of multistakeholderism where all stakeholder views and scientific evidence receive attention. The report insisted that global internet governance should be collaborative, decentralized, open and evidence based. It is acknowledged, however, that geopolitical considerations mean that, ultimately, internet governance is about 'the distribution of power in the political realm' (GCIG 2016, np), a realm which is characterized by power asymmetries. Another illustration of the privileging of constituted authority in the digital environment is the European Commission's use of its High-Level Group of Scientific Advisors to gather scientific opinions on cybersecurity issues. In this instance, the science advisors argued for principles such as transparency, care towards customers and shared responsibility among stakeholders, consistent with a multistakeholder model of governance (EC 2017b). However, the primary locus of decision-making authority remains the hierarchical apparatus of the European Union and the dominant paradigm of science was privileged.

A pluralist notion of power informs many of these kinds of initiatives to devise novel forms of governance insofar as multiple stakeholders are admitted to processes of consultation. This opens up a space for researchers who adhere to the hierarchical standards of science to offer their evidence and it occasionally admits those who work with critical theories in the constructivist tradition of science. However, decision making authority persists in resting with the state and/or the digital technology and service companies. In practice, power is reconstituted asymmetrically and hierarchies persist, consistent with the constituted authority approach, notwithstanding some evidence of the presence of elements of the adaptive authority model. Emerging forms of multistakeholder governance retaining features of constituted authority are especially common when choices about values have to be made or economic resources are at stake although, on occasion, there may be some convergence in values in areas such as a common goal of limiting infringements of fundamental human rights (DeNardis 2020).

The transition from the contract between ICANN (a multistakeholder governing authority institution aspiring to adaptive authority) and the US Department of Commerce (National Telecommunications & Information Administration—NTIA) (a constituted authority institution of the state) is an illustration of this. The contract enabling Internet Assigned Numbers Authority (IANA) functions to be performed expired in 2016. The successful transition to the multistakeholder organization, ICANN, to enable it to perform these functions under its own authority was regarded as an instance of assuring the accountability of all stakeholders including businesses, academics, technical experts, civil society (and researchers) and government stakeholders. The process of transition involved the constituted authority of the courts, however. Before the transfer date, President-Elect Trump and Republican Senator Ted Cruz indicated their opposition to the proposed arrangement, claiming that the United States government needed to retain its contractual interest to preserve the stability of the internet and its governance (Eggerton 2016; ICANN 2016). Their argument was dismissed by the court, and the transition was completed. However, the outcome might instead have favored the constituted authority of the state, highlighting the persistence of struggles for power in the internet governance domain between state and civil society actors.

In other areas of the digital environment, governance is less by the state, and more by the private sector, which exercises its authority to determine the outcomes and consequences of scientific and technological innovation. For example, internet protocols are often treated as technical issues, and decisions about implementation are taken by the private sector with little or no oversight by multistakeholder institutions. The Domain Name Service RPZ (Response Policy Zone) technology is an illustration. This technology allows an internet name server administrator to overlay customized information on top of the global Domain Name System (DNS), enabling responses to queries that differ from those that would otherwise occur (Vixie 2010). This has been deemed to be outside the remit of the multistakeholder organization ICANN, since ICANN does not have a mandate to intrude into matters which concern content, consistent with the principle of an open internet. A draft Internet Engineering Task Force document states that this technology 'merely formalizes and facilitates modifying DNS data on its way from DNS authority servers to clients' (Vixie and Schryver 2017). Yet this technology can be used by governments and companies to introduce content blocking by stopping access to certain servers, redirecting users to online walled gardens, or defending internet servers against cyber-attacks. The technology is being provided in the market by companies such as DissectCyber, Spamhaus, and ThreatStop. It has a socio-technical ordering or governing effect with the potential to be used to fragment the internet and it may be used to abrogate the rights of internet users to access content. The results of the dominant paradigm of science and technological innovation feed, in this case, into a closed system of governance with results supporting the private sector's interest in fragmenting the internet as well as state interests in controlling access to content. Treated as an instance of science diplomacy, albeit with the private sector in the lead, this example is indicative of the use of science to achieve normative outcomes that support prevailing power asymmetries.

In view of the variations in scientific norms and in establishing authoritative governance arrangements, the extent to which specific internet governance arrangements are effective in reconciling or accommodating conflicting stakeholder interests must be an empirical question (Mueller 2010; Cammaerts and Mansell 2020). Research on multistakeholder governance structures and processes indicates that it favors the interests of corporate actors due to their power in the market and that, in fact, it grants 'private interests legitimacy in public policymaking next to elected governments in the process' (Sarikakis 2012a, 151). Although internet governance arrangements may more often be characterized as heterarchical (Brousseau, Marzouki, and Méadel 2012) and be more reflexive (Hofmann, Katzenback, and Gollatz 2017) than in other sectors, this does not guarantee that an uncontested deliberative space will be created. This has implications for how research in the internet governance field is likely to be received in the context of science or digital diplomacy.

INTERNET GOVERNANCE RESEARCHER CONTRIBUTIONS TO DIPLOMACY

When internet governance researchers working in the constructivist traditions of the social sciences engage in science or digital diplomacy, they confront the prevailing norms of constituted authority. As argued earlier in this chapter, the dominant institutional norms of science diplomacy are essentially hierarchical in the way they confer status on researchers and their evidence. When researchers working in the internet governance field embrace normative positions with respect to socio-technical controversies involving the digital environment such as fundamental rights to freedom of expression, privacy, and access to digital information, state and corporate actors will seek to 'route around' them by ensuring that the authority of evidence, decisions and their enforcement continues to reside with the state or the private sector (Butt 2016; Mansell and Steinmueller 2022). Understanding how this happens requires research consistent with adaptive authority norms, that is, constructivist approaches, to lay bare 'the micro practices of governance as mechanisms of distributed, semi-formal or reflexive coordination, private ordering' (Epstein, Katzenback, and Musiani 2016, 4) or, as Milan and Ten Oever (2017) put it, to reveal the benefits of internet governance arrangements that are designed as a 'a normative "system of systems"'. These approaches aim to be consistent with adaptive authority models and with attempts to achieve equity among all of the stakeholders.

Understanding how more informal adaptive models of governance operate requires an examination of the performative agency of civil society stakeholders, including the scientific experts, who participate in diplomacy when they aim to influence internet or digital governance arrangements and outcomes. Revealing the norms and power dynamics of 'governance by social media' or 'governance by infrastructure' (DeNardis and Hackl 2015; DeNardis and Musiani 2016), can be used to highlight the corporate interests that typically are accorded dominant authoritative status in governance decisions. Researchers working in these traditions are providing evidence of how novel outcomes can emerge from the interactions of multiple actors with heterogeneous interests. Much of this work examines multistakeholder institutions such as ICANN or the discourses of stakeholders involved in, for example, the WSIS or the IGF (Epstein et al. 2016; Pohle 2016). However, these institutions that play a role in shaping internet governance have a greater affinity to elements of adaptive authority than do other institutions in the digital environment such as regulatory agencies which are increasingly active in governing digital technology systems (EC 2020; Mansell and Steinmueller 2020). van Eeten and Mueller (2013, 730) note the need for research that provides insight into 'environments with low formalization, heterogeneous organizational forms, large number of actors and massively distributed authority and decision-making power'; environments consistent with adaptative authority. However, governance in the digital environment continues to occur in settings where the norms of constituted authority are privileged. The result is that the asymmetrical power of states and companies is reconstituted, notwithstanding the appropriation of the term 'multistakeholder' by some these actors. As Chenou (2010, 26) observes 'the marks of a pro-business, technocratic, a-political and neoliberal power elite can still be found today in the debates on the future of Internet governance'.

Researchers in the field of internet governance who engage in science diplomacy also face the pitfall of co-optation. Franklin (2013b, 36) points out that engagement runs the risk of being 'disciplined into a post-Westphalian frame of institutional power'. When efforts are made to preserve constituted authority (of state or company) in contexts where the internet governance research community seeks to uphold citizens' fundamental rights, state and corporate perceptions of threat and vulnerability are heightened (Cammaerts and Mansell 2020). Science or digital diplomacy can become little more than an effort to persuade others to empathize with state policies since 'even an ethically informed mode of engagement cannot sidestep power asymmetries' (Comor and Bean 2012, 215). Diplomacy, drawing instead upon the results of internet governance research in the adaptive authority mode of science may be more open and transparent as a result of multistakeholder governance forums. However, although some digital companies are encouraging a more open circulation of information and the involvement of new voices (Dutton 2017), the potential for conflict between stakeholders favoring constituted authority and those favoring heterarchical adaptive authority institutions of governance is substantial.

This does not mean that the outcome of struggles for authority inevitably will favor states and the corporate actors. Parkinson (1958, 8) insisted that 'there is, in fact, no historical reason for supposing that our present systems of governance are other than quite temporary expedients'. They are always contested. And, following Slaughter (2004), in a 'disaggregated world order', fluctuations in the roles of states and intergovernmental organizations are common (Schemeil 2013). These fluctuations should be expected with respect to the authoritative status of digital technology companies and concerned civil society actors, including the scholarly community, and adversarial conflict can be generative of new sites of hegemony (Cammaerts and Mansell 2020). Greater participation by internet governance researchers in science or digital diplomacy entails their immersion in agonistic confrontations where stakeholders with oppositional goals are operating in a contested adversarial space (Mouffe 2013). In the case of digital socio-technical controversies, this dynamic space can yield re-articulations of power that generate new, albeit temporary, hegemonies in which civil society stakeholders may be able to establish their authoritative status and achieve changes consistent with citizens' interests.

CONCLUSION

Internet governance researchers who argue for the protection of citizens' fundamental rights have opportunities to bring about change when they engage in science or digital diplomacy with other stakeholders in the struggle to establish a hegemonic position that differs from prevailing interests of the dominant digital platform companies and of states. The prevailing hegemony of hierarchical science and constituted authority-styled governance institutions tends to exclude or downplay the significance of the views of those (such as practitioners, those who contribute to 'research in the wild', and researchers aligned with the constructivist tradition) who are not accredited in accordance with the prevailing paradigm of science. Nevertheless, contestations over the authoritative status of stakeholders and research evidence have the potential to mobilize momentum towards changes in internet and digital governance such that fundamental rights of citizens can be privileged over or alongside the rights of other stakeholders.

The engagement in diplomacy by internet governance researchers working in the constructivist social science tradition brings them into close proximity with the norms and values of constituted authority institutions of the state, the corporate world, and the hierarchical science norms with the risk that their evidence will be devalued. The evidence they provide may help, however, to persuade those whose preference is for constituted authority governance arrangements and outcomes that the digital world is a complex evolving system. It is co-constituted by multiple actors, albeit within a structure of asymmetrical power relations in the capitalist economy. State and corporate representatives who engage in science or digital diplomacy to address challenging digital controversies might come to better understand why controlling the internet and the wider digital environment through hierarchical norms and practices is ineffective as a result of their encounters with internet governance researchers whose evidence probes and explains this complexity. Their engagement with researchers bringing insight into the viability of adaptive authority governance arrangements is likely to lead to an accumulation of evidence providing greater confidence in the outcomes of adversarial adaptive authority governance processes undertaken by democratic multistakeholder institutions. The dynamics of these interactions are yielding new hegemonies within the sites of internet governance and, in this way, as Williams (1983, 268) argued, 'once the inevitabilities are challenged, we begin gathering our resources for a journey of hope'. In this instance, the hope must be for the construction of a digital world through the soft power of science or digital diplomacy that, ultimately, develops in the interests of citizens in equity, openness, and inclusiveness.

The norms of a non-ideological and universally authoritative science that seems to yield insight into optimal ways of governing the digital world in line with constituted authority governance norms will persist. Conflict is likely to grow, however, as diplomats turn to the internet governance research community for assistance in tackling global digital challenges. Diplomats will expect scientifically validated research consistent with the predominant notion of a 'completed', value free, science and this will intensify agonistic relationships among the stakeholders. However, it is also likely to generate new opportunities for the emergence of novel governance structures and practices some of which may be consistent with adaptive authority approaches to governance.

In other areas where researchers address socio-technical controversies in fields such as climate change or disease prevention where conflict arises, constructivist social science researchers have proven themselves to be influential when they adhere to 'technologies of humility' in their promotion of discourses favorable to democratic deliberation and equitable outcomes (Jasanoff 2003, 227). The pitfalls associated with co-optation through engagement in diplomacy as a result of a normalization to the hierarchical values of constituted authority as a 'standard' pattern are considerable, but they can be resisted through the reflexive practice of researchers and other civil society members, consistent with the tenets of the constructivist tradition. When they are resisted effectively, there is a greater chance that evidence-based approaches to digital governance can become supportive of policies and practices that provide stronger guarantees of the fundamental rights of citizens.

Chapter Six

Crafting Science Diplomacy in Comparative Perspective

The Case of US Internet Governance

Nanette S. Levinson

E. William Colglazier and Elizabeth Lyons (2014) write that the 'STI' (Science, Technology, and Innovation) enterprise will need to adapt to new opportunities and changes in the current landscape of global science. To be most effective, the response should include embracing a strategy of international STI research cooperation and utilizing STI knowledge strategically by looking *out, up, around, and forward.* This can empower the US STI enterprise, especially its decentralized academic components, to engage globally (and) facilitate strategic international STI cooperation. They go on to argue that such efforts are essential to the national interest. This chapter analyzes the emergent field of science diplomacy and related diplomacy in comparative perspective, paying particular attention to relevant cross-cultural communication and public diplomacy research and writings as well as to the public and science diplomacy practices of the United States.

It recognizes that there is very little published work on the case of internet governance in the context of science diplomacy or public diplomacy with a focus on the United States. Rather, the incipient science diplomacy literature focuses on, for example, health or energy or even foreign policy and the role of science diplomacy. Recent literature does begin to treat aspects of internet governance in the context of science diplomacy, especially in the case of cybersecurity and artificial intelligence. See, for example, Tanczer, Brass, and Carr (2018) who analyze computer security and incident response teams (CSIRTS) as actors in science diplomacy, utilizing their formal and informal networks to share knowledge cross-nationally. There is also recent work (Ulnicane et al., 2021) using artificial intelligence as a locus for science diplomacy and international cooperation initiatives. This chapter builds on this work and examines the case of internet governance during the past decade with a particular science and cultural diplomacy lens, highlighting studies from around the world. It also argues that to analyze most effectively internet governance in the context of diplomacy, there is a need to examine cultural and knowledge transfer patterns amongst its multistakeholder setting involving scientists, policymakers, diplomats, and related networks including that of the private sector.

Diplomacy as a concept has existed since ancient times. Traditionally, diplomacy has been a government-to-government phenomenon and refers in modern days to the actions of foreign ministries to promote the policies of their own country to other governments and their foreign ministries and leaders. With the rise of communications technologies and especially the internet, diplomacy has taken a new turn, sometimes successful and other times not. It simply has added new media to the conduct of diplomacy as well as a science and technology thrust to diplomacy. The conduct of diplomacy still emanates primarily from a government agency. However, with the addition of new media, the focus of diplomacy broadens dramatically to easily include foreign publics as well as foreign ministries. The focus has amplified to encompass even more directly influencing people as well as governments. This turn to 'public diplomacy' parallels the growth of the concept of soft vs. hard power (Nye 2008; McClory and Harvey 2016), emanating from the end of the Cold War. It also parallels the emergence of internet-facilitated communication including social media. While some may argue that public diplomacy dates back to mass media including radio, new and internet-related media have transformed 'diplomacy places' as this author terms them and discusses in the next section of the chapter.

DIPLOMACY PLACES

Two well-recognized subsets of diplomacy are cultural diplomacy and science diplomacy; these diplomacy domains have existed for decades and continue today. Foreign governments often send to the United States their top orchestras or ballet corps or art exhibits or even sports figures or teams to link to the people of the US. Similarly, the US sends its leading music, cultural, and sports teams or figures to other countries to build goodwill and bridge across cultures. Regarding science diplomacy, scientists have had cross-national ties well before the advent of the internet and the rise of social media. However, these ties related almost exclusively to the transfer of knowledge within specific scientific fields.

A US governmental focus on science diplomacy, though, is a more recent development. As Domingues and Neto (2017) point out, science diplomacy is now a tool of soft power, especially in the context of the United States.

Similarly, Moreno et al. (2017) highlight the roles of science diplomacy but in the context of Spanish public administration. These authors argue that the Spanish government is utilizing Spanish scientists located around the world to promote Spain's national interests. The idea of diasporas and the promotion of national interests through these scientific or even business diasporas such as that described in the case of the Spanish government is taking hold.

An additional but less well-known subset stems from the field of peace and conflict resolution studies and originates outside of formal government agencies: Track 2 Diplomacy (Montville 1991b; 2006). Track 2 diplomacy involves individual groups outside of formal government agencies coming together to meet and develop consensus in the context of peacebuilding, especially in conflict zones. This type of diplomacy can also hold potential for analyzing internet governance as public diplomacy. It resonates well with the involvement of stakeholders in addition to governments and international organizations in internet governance multistakeholderism processes and practices. Yet there has been scant literature that compares Track 2 diplomacy in peacebuilding with internet governance processes and practices. Such a comparison could hold important implications for viewing the field of internet governance through the lens of diplomacy writ large.

Similarly, governments including the US government are recognizing potential diplomatic roles of the increasingly multinational private sector. Indeed, in the case of internet governance and especially cybersecurity, a highly complex and science and technology-based local and global arena or ecosystem, the private sector possesses the potential to serve as a key actor in diplomacy efforts and in furthering national interests. An entire industry has grown up around the Internet, both within the United States and throughout the world. Recent work on 'Green Public Diplomacy' highlights US diplomatic efforts with China during 2008–2014. The authors of this study (Yang, Wang, and Wang 2017) find that the impetus for diplomatic efforts related to the environment shifted by 2014 from the two governments involved (US and China) to businesses and NGOs. We are now seeing a similar shift in the realm of cybersecurity at the same time as we are seeing a reinvigorated role for national interests and national governments.

Internet governance today, as messy and complex as it is, involves interactions among civil society, technical, and private sector organizations as well as nation-state and international organizations. Thus, one can consider research on track 2 diplomacy and track 1.5 diplomacy (wherein governments AND non-state actors conduct diplomacy in conflict situations) as a powerful way to understand multistakeholderism and the myriad networks both formal and informal in the national and global internet governance policy spaces as a potential tool for public diplomacy. One of the growth areas for public diplomacy is science diplomacy, as discussed earlier. It builds on the infor-mal connections that have existed among scientists globally and the complex challenges in science (including internet governance/cybersecurity/related emerging technologies arenas) that provide impetus for both cooperation and competition. The presence today of internet-facilitated cross-national connec-tions makes it even easier for informal networks of scientists and technical experts such as those computer scientists and engineers involved in internet infrastructure and in technical standard setting to be forged and maintained. In the internet governance ecosystem, standard setting organizations such as the IETF, thrive on the participation of internet technical experts (and their companies) both within and across nations.

Yet scientists' communication patterns are traditionally different from those of diplomats. Linkov et al. (2016) argue that the networks and accom-panying knowledge transfer patterns of scientists and those of diplomats are quite differentiated; using both strengthens diplomatic efforts overall. Applying Linkov's work to the internet governance arena highlights the dif-ferent occupational cultures and communication patterns as well as technical capacities of computer scientists, civil society representatives, business-people, international organization bureaucrats, and diplomats. Such nuanced differences present nuanced obstacles (and opportunities) for the conduct of internet governance as science diplomacy. The recognition of these nuances also helps to facilitate science diplomacy to promote a nation-state govern-ment's national interest.

Three subsets of science diplomacy emerging in recent years also link to major global challenges and accompanying uncertainties as well as techno-logical, social, and cultural factors. These are: medicine, health, and espe-cially in view of the COVID-19 pandemic, vaccine diplomacy (Hotez 2014; Linkov et al. 2016). Hotez (2014) cites the medical diplomacy conducted by Cuba over a fifty-year period with regard to other Latin American nations. He calls for a joint vaccine development initiative by the US and Cuba as well as with other Latin American nations where there is a dire need for vaccines against debilitating and less known tropical diseases. Arguing that such efforts build positive scientist-to-scientist ties even among nations with differing foreign policies and hostile stances, Hotez notes that such ties and the resultant success of such vaccines can lead to positive public attitudes. This is, indeed, public diplomacy with a focus on capacity building. In light of the COVID-19 pandemic, Cuba, China, and other countries including the United States are providing vaccines and vaccine knowledge to developing nations. These efforts can also serve as benchmarks for the field of internet governance wherein technical experts from one country or a group of coun-tries can participate in capacity-building efforts in other countries, thus also forging cross-national knowledge transfer ties and fostering positive public attitudes.

There is a final and often less studied type of diplomacy this chapter needs to highlight. It is the diplomacy carried out by international organizations, historically considered as merely facilitating mechanisms among member nation-states and not actors in their own right (Levinson and Marzouki 2016). Today international organizations are proactive in crafting ideas to promote diplomatic efforts in science and technology domains writ large. See, for example, UNESCO's efforts to craft and promote the concept of 'internet universality' in the field of internet governance (Levinson and Marzouki 2016). This work involves not only member states but extends to nongovernmental organizations and to academe.

POLICY SPACES AND DIPLOMACY PLACES

In the United States, the public administration/technology policy scholar W. H. Lambright (1976) coined the term 'policy space' to capture which US government agency had primary responsibility for which policy domains. The US Department of State has had and continues to occupy the policy space for traditional diplomacy and for public diplomacy. Recent work on science diplomacy also illustrates vividly the central role of a Ministry of Foreign Affairs in the case of Spain and public diplomacy (Moreno et al. 2017) or the role of the Department of State in the US. What is particularly fascinating here is the Moreno et al. table listing those involved in public/science diplomacy in the context of their study: it includes the 'Special Ambassador for Cybersecurity' alongside of 'Institutos Cervantes', a government-funded organization promoting culture and language.

Turning to the United States overall, policy space is a much more complex issue when we move to the topic of internet governance diplomacy, the focus of this chapter. The boundaries for policy spaces have become fuzzier and more porous. Regarding science diplomacy itself (encompassing computer science research), the US National Science Foundation plays a key role, as does the US National Institutes of Health with regard to health policy. Additionally, the US National Academy of Sciences, along with sister academies in medicine and engineering, for example, are key science diplomacy players. In sum, the policy spaces for science diplomacy today in the US cover a plethora of government agencies and extend to incorporate nongovernmental players and networks as noted below. However, focusing on internet governance policy and especially cybersecurity poses even greater challenges for defining these policy spaces, especially where a US government agency also has responsibility for national security as in the case of the US Department of Homeland Security or the US Department of Defense. Even without a focus on the national security components of things cyber, the US government policy space for internet governance remains complex, with the US Department of Commerce and the US Department of State playing key roles.

Another more recent and catalytic space for science diplomacy is evident in the US in the American Association for the Advancement of Science (AAAS), a nongovernmental association of scientists from myriad fields. AAAS has, indeed, a Center for Science Diplomacy, founded in 2008, with a goal of 'using science to build bridges between countries and to promote scientific cooperation as an essential element of foreign policy by raising the profile of science diplomacy, creating a forum for thought and analysis, and initiating bilateral activities'.[1] It bestows an annual award for science diplomacy and since 2012 has published a journal entitled *Science and Diplomacy*. The 2016 AAAS award citation commends the 2016 annual award winner, the Honorable Grace Pandor from South Africa, for her successful efforts to incorporate science across a range of South Africa's governmental bureaus. Note that this award itself exemplifies a facet of public diplomacy, here a US nongovernmental organization recognizing the efforts of a South African leader, promoting the notion of science in diplomacy, and building good will. This 2016 award to the Honorable Grace Naledi Mandisa Pandor, South Africa's Minister of Science and Technology, primarily illustrates one side of the science diplomacy coin: science in diplomacy through strengthening scientific expertise and knowledge in government agencies. However, Pandor was also cited for her work in promoting collaborative partnerships across nations that exemplifies diplomacy through science.

These complex policy spaces for both science in diplomacy and diplomacy for science contain equally complex knowledge transfer patterns and link to other relevant policies such as a government's innovation, competition, or security policies. Focusing on knowledge transfer, there are questions regarding whether the various types of science diplomacy are most effective as top down or as bottom-up efforts. Linkov et al. (2016) argue that government agencies' top-down approaches are more effective than bottom-up approaches. Contrastingly, the earlier cited February 2017 issue of *Science and Diplomacy* contains an article on Spanish science diplomacy and its successful collaborative and bottom-up approach, especially its focus on linking the Spanish diaspora. Rigorous research is needed to analyze in detail the benefits and drawbacks of differing approaches and the need to incorporate growing informal non-state actor or mixed state/non-state actor cross-national networks. There may even be cultural aspects of different approaches; and these would need to be considered as well in further research studies. Another arena for additional research relates to what I prefer to call diplomacy through science (or medicine or health or vaccines) and goes beyond the top down vs. bottom-up dichotomy. It requires an understanding of what I have previously called co-creation processes (Levinson 2015). I will discuss these collaborative processes especially in the context of science diplomacy and internet governance below.

A final trend in terms of science diplomacy is the 2015 call by the US National Academy of Sciences, Engineering and Medicine (National Research Council 2015) for 'widespread involvement of public and private sector organizations throughout the country (to) play important diplomatic roles in many areas involving S&T considerations such as Internet governance, foreign trade, global scientific research programs, and humanitarian assistance.' Similarly, a recent head of the American Association for the Advancement of Science (AAAS), Rush Holt, emphasized the need for the US State Department to prioritize science and technology and to replace the departing Science and Technology Advisor expeditiously (Holt 2017). Holt argued cogently for the benefits of such a role. As of 2021, the US Department of State has an Acting Science and Technology Advisor in the permanent Office of the Science and Technology Advisor.

INTERNET GOVERNANCE POLICY SPACES

Similar to medical, health, and even vaccine diplomacy is the fractal policy arena or policy spaces occupied by internet governance. Internet governance is both local and global, national and international; it occupies policy spaces that encompass in messy ways national government agencies (and, indeed, a changing panoply of these across time!), international organizations, civil society organizations, private sector organizations, and even technical organizations which some may classify within civil society.

The 2005 World Summit on the Information Society statement (ITU 2005) includes relevant sections related to internet governance and reads as follows:

29 . . . the Internet has evolved into a global facility available to the public and its governance should constitute a core issue of the Information Society agenda. The international management of the Internet should be multilateral, transparent and democratic, with the full involvement of governments, the private sector, civil society and international organizations. It should ensure an equitable distribution of resources, facilitate access for all and ensure a stable and secure functioning of the Internet, taking into account multilingualism . . . (and) 34. A

working definition of Internet governance is the development and application by governments, the private sector and civil society, in their respective roles, of shared principles, norms, rules, decision-making procedures, and programmes that shape the evolution and use of the Internet.

This statement outlines the perimeters of the policy spaces and allows both for the wide range of policy issues and for the numerous actors/organizations involved in internet governance today. It also serves as a foundation for the term 'multistakeholder' as it applies to the internet governance policy spaces and especially to the Internet Governance Forum (IGF) created as an outcome of the World Summit on the Information Society and codified in the 2005 statement discussed above. The purpose of this organizational innovation, the Internet Governance Forum (IGF), was to serve as a 'multistakeholder' dialogue entity. Stakeholders include the technical community. (Please note that this author has traced the use of 'multistakeholder' to another, earlier United Nations policy space, that of environmental governance, thus demonstrating mimetic isomorphism and knowledge transfer across increasingly interconnected global governance related science policy spaces).

Moving to the United States in historical context, internet policies began in a simple governmental policy space, that of one government agency, the US Department of Defense's, Defense Advanced Research Projects Agency (DARPA). As time passed and governmental actors recognized the commercial potential of the internet and even its global potential, the primary policy space moved to the US Department of Commerce. Increasingly there was recognition within the US government that there was also a diplomacy side—an international side—to internet governance policies, thus including in the US policy space, the US Department of State.

This understanding began to take hold as far back as 1997 when under the administration of former US President Clinton, the US Department of Commerce was charged, in recognition of the international dimensions of privatizing the domain name system of the Internet, with facilitating international participation in its governance. It did so through inviting and receiving comments in preparation for its 1998 Green Paper. This, too, added to the geometric complexity of internet governance policy space in the United States. For the first time, the term "policy space" needed to extend beyond the US Department of Defense and the US Department of Commerce, to other agencies, other countries, and even nongovernmental actors.

Today, the increasing national governmental concern with cybersecurity brings back, in a more central way, the US Department of Defense and also, of course, the newer Department of Homeland Security. While it is straightforward to identify the players in the US government policy space for internet governance, there is less research on the informal power panoplies (possibly changing according to the nature of a policy issue) within this geometric configuration. Note that these power panoplies may fluctuate in the United States, according to the platform of the political party in power at any given time.

One of the features of internet governance as noted in its 2005 definition that lists who is involved is the notion of multistakeholderism, as noted earlier in this Chapter. This refers to the involvement, depending upon the type of policy issue, of public sector, private sector, civil society and technical actors. Similar to environmental governance, health governance, and other science- and technology-related arenas, internet governance often involves a multistakeholder policy space, some of which technical experts occupy. This refers to who is present but does not specifically delineate the power or influence or exact role of each stakeholder. It also does not identify what officially constitutes a stakeholder in multistakeholder settings. Note also that the earlier-noted 2005 statement refers to the respective roles of the different actors, invoking subtly reminders regarding power(s) of nation-state governments vis-a-vis other actors who are newer to the diplomacy 'table'.

INTERNET GOVERNANCE AND SCIENCE DIPLOMACY: THE WAYS FORWARD

Today, within internet governance and its multistakeholder context, as fuzzy as that might be, there is a panoply of various types of actor and organizations at local, national, international, and cross-sector levels, each with various motivations, powers, and even cultures. This equation or ecosystem (Levinson and Marzouki 2016) parallels well the most recent focus on science diplomacy—neither a top-down nor bottom-up approach. Rather the approach is more multiplex in nature. It allows for the following elements, in the context of internet governance with its kaleidoscopic policy spaces:

- Standards setting organizations comprised of technical experts to do their work (but to what extent, if any, does a national interest influence any participants?)
- Multistakeholder institutions such as the IGF (with recent efforts at the United Nations to strengthen the IGF or similar mechanism)
- Governmental and international organizations crafting policy in their spaces, but most expounding a multistakeholder approach.
- The Internet Governance Forum with its multistakeholder composition as a locus for cross-sector and cross-cultural dialogue

- International organizations that have included internet governance as related to their missions and their informal and formal networks with non-state actors.
- The cybersecurity agencies within and across countries, sometimes guided by non-state actor consultation.

Just the listing of these elements conveys a complex ecosystem. It is within this multistakeholder ecosystem that knowledge transfer does or does not take place. Additionally, the national, ethnic, organizational, and even occupational cultures present shape (both positively and negatively) the flow of knowledge and thus influence outcomes and impacts of these patterns. Here is where work on co-processes (Levinson 2015) becomes especially relevant. Is there a co-process situation enacted in a given multistakeholder setting such as internet governance diplomacy? How does it actually work? What is the result of these co-processes and can there be both collaboration and competition in such processes as there has been in the case of formal partnerships and alliances across sectorial or even national boundaries? The next step is to gather data regarding the knowledge transfer patterns (formal and informal) among stakeholders with particular attention to cross-cultural communication patterns. Research that particularly addresses the knowledge flows across stakeholder groupings, each of which has its own culture, and each of which may be embedded in a national culture, can elucidate the conduct and, indeed, the impacts, of cross-stakeholder diplomacy-related knowledge flows (see, for example, Stone, 2019 or LeGrand and Stone, 2018).

CONCLUSION

One of the fascinating developments in the realm of diplomacy is the amazing growth of myriad diplomacies over time: cultural, sports, public, science, medicine, health, vaccine, and now, internet policy. There is not much rigorous data on whether these diplomacies are effective or not. In fact, viewing internet governance, there are some who have argued that the United States supports multistakeholderism because, in the end, it is in their national interest, and, indeed, is a way to keep those governments that are authoritarian at bay in the internet governance policy arena.

The last decade has seen a growth in the idea of cooperation and competition within partnerships and alliances across sectors and across national boundaries. This mimetic isomorphism foreshadows multistakeholderism across organizational types, national homes, sectorial types, and even cultures. We have, as noted above, scant in-depth, holistic, and longitudinal research that captures these interactions in these nuanced internet policy spaces.

What is needed now is rigorous research, both qualitative and quantitative, to examine the detailed and nuanced operations of all types of science and public diplomacies (including internet governance), with a focus also on co-process outcomes, barriers, and impacts. Additionally, there is a need to examine specifically the roles of universities and think tanks in these complex multilevel policy spaces that today also can constitute science and public diplomacy spaces as well. Finally, power in all of its various meanings, cannot be overlooked. To what extent do governments, as was evident in the example of Spanish public diplomacy via its scientific diaspora, play a role in catalyzing other stakeholder roles in carrying out an individual government's national interest? While the Yang, Wang, and Wang (2017) study tracked a movement away from government-to-government diplomacy in the environmental governance policy space, there may be today a shift back toward more direct governmental involvement in the internet governance policy space with the advent of increasing cross-national cybersecurity concerns. At the same time, with ever-converging internet-related technologies, social media, cross-cultural communication, and public attitudes need to be considered as components of internet governance roles with regard to policy and practice.

NOTES

1. See https://www.aaas.org/program/center-science-diplomacy.

Chapter Seven

Modes of Internet Governance as Science Diplomacy

What Might the EU Learn from the US Cybersecurity Policy?

Francesco Amoretti and Domenico Fracchiolla

A COMPREHENSIVE DEFINITION OF SCIENCE DIPLOMACY

The COVID-19 pandemic relaunched dramatically the relationship between science and diplomacy at international level. The assertive posturing of US and Chinese technology companies during the pandemic seems to confirm the provocative conviction of US political scientists Henry Farrell and Abraham L. Newman (2019) that interdependence is not only a promise but also a danger. They argue that states are currently finding themselves overwhelmed by private actors controlling global network, forums, and supply chains as a source of power and potential weapons against internal and international opponents, on a number of issues, from innovative technology to needed medicines, digital communications and network infrastructure, as well as access to the global financial and monetary system.

To counterbalance that catastrophic interpretation, the global vision of those crises has been strengthened, highlighting the importance of the close relation between science and diplomacy as the main reference to tackle global challenges, capable of endangering the whole international community. To explicate the seminal and productive long-term history of intersections and reciprocal influences of the relation between science and diplomacy there are a number of reasons that will be discussed. However, that relationship is also a story laden with much rhetoric. As the US House Committee on International Relations stated in 1977:

> Science and technology have affected changes in the substantive tasks of foreign policy, in the methodology of diplomacy, in the management of information on

which diplomacy is based, in the intellectual training of diplomats, in the range
of present options of negotiations, and in the [. . .] prospects of future evolution
of diplomatic, foreign policy objectives and the international political system.
(cit. in Dufour 2016)

Time has moved on, and science-based issues have become increasingly
important in the conduct of foreign policy, heightening the need for policy
makers to develop and implement science diplomacy strategies (Cooper,
Hocking, and Maley 2008). Despite this long and rich history, science diplomacy, *as a concept*, is rather new. It dates back to 2010, when the Royal Society and the American Association for the Advancement of Science identified
three main type of activities or analytical dimensions, that is, science in diplomacy, diplomacy for science, and science for diplomacy. Their perspective,
shared by other scholars, has strongly influenced the debate on the academic
and institutional level. The underlying idea is that countries are seeking to
achieve some or all the three 'Es' of science diplomacy: 'expressing national
power or influence, equipping decision makers with information to support
policy, and enhancing bilateral and multilateral relations' (Turekian 2012).[1]

In recent years, some initiatives gave a new impulse to that perspective,
giving to it more authority thanks to the growing involvement of the international scientific community. In that vein, the 'Madrid Declaration on Science
Diplomacy' (S4D4C 2019) is a strategic document signed by a group of
experts, confirming the high value of science diplomacy to inform a new and
more effective governance structure for statecraft. The document sustains the
value of science diplomacy to support the universal scientific and democratic
principles, including science and technology as key dimensions of foreign
policy and international relations at different political levels, realizing a more
productive and sustainable international relations.

In addition to this, the Declaration states the great potential contribution
of science diplomacy to address global challenges and it seals the works and
the interpretations of scholars that have already elaborated concepts on that
regards.

For example, for Van Langenhove (2016a), science diplomacy has the
potential to be a 'change agent', while Turekian, Gluckman, Kishi, and
Grimes (2018) posits that science diplomacy as a soft power tool of potential
strategic relevance to tackle the current global challenges referring to the
loss of the former traditional hegemony of the Western world. For Copeland
(2016) 'Nevertheless, SD is important and is becoming more so in an increasingly heteropolar world order where the vectors of power and influence are
characterized more by difference than by similarity and S&T based challenges are multiplying.' In line with the pragmatic approach by Turekian,
Gluckman, Kishi, and Grimes (2018), 'science diplomacy is coming to the

fore as a formidable dimension of interstate power relations. As the challenges of the world increasingly transcend borders, so too have researchers and innovators forged international coalitions to resolve global pathologies' (Legrand and Stone 2018).

To conclude, the Declaration underlines the importance of SD to exert the tool of the independence of science, to benefit from capacity-building activities, to foster the inclusion and the evidence informed in foreign affairs policies (S4D4C 2019).

It is interesting to note that the description of the institutionalization of the science-policy relationship, broadly understood, 'does not consider the Internet Governance as a global challenge'.[2]

We may suppose that this is due to three main, interrelated reasons. First: unlike policy issues such as environmental change, human health, sustainable and inclusive development, Internet governance has not yet reached a comparable level of institutionalization. Commenting over the first global conference on Internet governance hosted in Brazil in 2014, Josef Nye wrote: 'At the end of the conference, there was still no consensus on global cybergovernance . . . In a sense, this is not surprising. After all [. . .], it has not been around for very long' (2014).[3] Second: The internet is a complex, fast-evolving, and all-encompassing global resource. This basically means that the scope of IG is a dynamic phenomenon, the borders of which is a matter of ongoing negotiation and may depend, as noted by Hofmann, Katzenback, and Gollatz (2014), 'on the observer as much as on the issue under investigation'. Compared to other global governance dynamics, IG processes are characterized by 'distributed collective action, overlapping authorities, competing rationalities and goals'. A lack of central control does not imply anarchy or absence of rules. To quote Brousseau, Marzouki, and Méadel (2012a, 17), Internet governance has, indeed, its own order, defined as a 'networked heterarchy' consisting of public and private organizations which interconnect with one another through various forms of 'mutual recognition and mutual legitimization'.

Finally: unlike Multilateralism 2.0, which presupposes that governance structures and processes increasingly marginalize states in favor of non-state actors, in the last few years IG is experiencing a crucial change that might be summarized in the following terms: States are back in cyberspace. We are aware that states have never left this competitive arena, even when it was at its peak the multistakeholder model. Such a claim, however, does not imply that we share a state-centric approach to cyberspace or that we adhere to that 'methodological nationalism' that has traditionally characterized mainstream security studies. Having gained this new centrality, states now have to update the set of means and modes of their security policies to respond to the threat of the complex security linkages existing across the world.

Cybersecurity aims to adapt national policies and the traditional general principle of security mechanism, like deterrence and collective defense, to the new reality of the twenty-first century, when the division between insider and outsider in nation-states has been challenged (Adamson, 2016). Our hypothesis is that the radical shift is due to the massive rise of cybersecurity issues on the political agenda of all the actors involved, and primarily of the national governments which are called to use new means and models of security policies.[4]

According to the OECD (2012) 'a comparative analysis of the new generation of national strategies reveals that the cybersecurity policymaking is at a turning point [. . .]: sovereignty considerations have become increasingly important'.[5] A chronological reconstruction of the evolution of Internet governance domain shows the relevance of cybersecurity and how it has been reformulated over the years, widening its scope to include different policies. On the other hand, this transformation does not mean that other policies are losing importance, though this might be a possibility in the future. It implies, on the contrary, that cybersecurity is the main asset and the final test of governmental policies dealing with cyberspace at the national as well as the international level—and at a bilateral and multilateral level alike.

In different words, the range of possible policies is essentially narrower, considering the cybersecurity dimension as a fundamental strategic policy for all the actors. From NGOs to international organizations (IO), from giant corporations of ICT to epistemic communities of academics as well as professionals and technocrats, a relevant reorganization of their agenda of actions and political and institutional initiatives in the cybersecurity domain has been registered in the last years. The relevance of this reorientation reached the attention of global media and international public opinion. This helps to explain why the international community is showing a growing interest towards the implication of Internet governance as science diplomacy, placing at the very top of its agenda the role of science and technology for supporting foreign policy also in this field. Though IG is not a policy coherently developed and with defined boundaries, like other domains of science diplomacy, IG as science diplomacy is becoming crucially relevant in political strategies of democratic as well as authoritarian governments, changing their structures and features as regards the reorganization of bureaucratic apparatuses and offices, reallocation of powers, financial resources and know-how.[6] Beside traditional policies, the intersection between IG and science diplomacy can therefore be deemed as the last frontier of the long-term process of intersection between scientific knowledge and political decision making at the national and international level[7].

FLUID POLICY IN IG STRATEGIC AXIS: THE POWER
OF CYBERSECURITY AS SCIENCE DIPLOMACY

Security concerns have been widening their prominence within the Internet governance debate and policy initiatives during the last decades. Therefore, cybersecurity as a policy issue has soon been institutionalized, but it has also affected other Internet governance issues, pervading the discourse on protocols and standards, on privacy and data protection, and even on other human rights. A relevant aspect of the discourse on cybersecurity is that it is serving as a common platform where the Western privatized model of governance is merging with governments' aims to strengthen their own role in the global Internet governance. This evolution is not a temporary phenomenon. It is, rather, a comprehensive dimension of policy being consolidated as a response to the challenges of cybersecurity. In particular, this development has intensified as a result of the threats posed by cyber-aggression or cyber-conflict, mainly fueled by the certainty that, on one hand, an increasing number of states use cyberspace to achieve strategic ends, and, on the other, that even non-state actors are exploiting cyber-tools potentialities for disruptive activities. In response to these threats, several multilateral organizations have proven useful in developing or coordinating cybersecurity policies. The work of the fifth UN Group of Governmental Experts (GGE), the launch of the Global Commission on Stability in Cyberspace (GCSC) at the Munich Security Conference in February 2017 and the proposal by Microsoft of a New Digital Geneva Convention in 2018[8] exemplify this reaction. Another example, at this level, is offered by the GFCE. The Global Forum on Cyber Expertise (GFCE) is a global platform for countries, international organizations, and private companies to exchange best practices and expertise on *cybercapacity building*. The aim is to identify successful policies, practices, and ideas and multiply these on a global level. Together with partners from NGOs, the tech community and academia, GFCE members develop practical initiatives to build cybercapacity.[9] These are important signs of a renewed interest in discussing the implementation of international rules to protect the civilian use of the internet. Recent developments would seem tied strongly to previous experiences, as in the case of UN GGE, or to continue the tradition of the multistakeholder approach, at least in lexical terms, by focusing on the use of the words in official documents. In truth, they represent a completely different and new scenario, one in which the states are rapidly and inevitably expanding their power in an attempt to gain more security.[10] Cybersecurity has become an integral part of governments' national defense and foreign and security policies and doctrines, contributing to the construction of a *new complex domain of policies*, where diplomatic talks and initiatives to achieve international agreements have become increasingly essential.[11] Overall, a secure, safe, and open cyberspace is impossible without the involvement and commitment of states. If the trend is to invest in relations to *normalize* cyberspace, focusing on measures for building confidence in cyberspace, it is useful to analyze national policies and strategies, starting from the paradigmatic experience of the United States. Our aim is to pinpoint their main features to better understand, in turn, the EU's current strategy.[12] Arguably, comparing the two experiences is the best way of offering suggestions for the future of the European cyberspace strategy.

THE US EXPERIENCE

The evolution of the US policies and strategies in cyberspace is important to understand how the latter has become the most important policy domain of Internet governance in the United States. Science diplomacy is a central component of America's twenty-first century agenda. The United States increasingly recognizes the vital role that science and technology might play in addressing major challenges, such as making the US economy more competitive, tackling global health issues, and dealing with climate change. The Department of State's first Quadrennial Diplomacy and Development Review highlights that 'science, engineering, technology, and innovation are the engines of modern society and a dominant force in globalization and international economic development'. The Department of State is committed to using US capabilities throughout the world to connect with scientists, entrepreneurs, and innovators for the mutual benefit of peoples. Today, the United States has formal science and technology agreements with over fifty countries. It is committed to finding new ways to work with other governments in science and technology, to conduct mutually beneficial joint research activities, and to advance the interests of the US science and technology community. Indeed, the overwhelming US dominance in scientific research in the last half of the twentieth century is being replaced by a more multipolar landscape of science, technology, and innovation, with the United States remaining a very strong force.

The starting point for understanding the situation before the Obama administration is the report on *The Pervasive Role of Science, Technology, and Health in Foreign Policy: Imperatives for the State* (National Research Council 1999) on the general topic of science diplomacy. Its most important outcome was the creation of the Office of the Science and Technology Adviser to the Secretary (STAS) incorporated in the legislation and endowed with the power of appointing advisers. The reorganization of important components of the department is a relevant issue of the report. An ad hoc committee was to assess the adequacy of the capabilities of the State Department to use effectively the nation's S&T assets in achieving US foreign policy objectives during the next decade.[13] The Presidential Decision Directive 63 (PDD-63), signed in May 1998, was centered on the defense of the US cybersystems as critical infrastructure. It argued that the US would have taken all the necessary measures to 'swiftly' eliminate any vulnerability to both physical and cyberattacks on its infrastructures, including, especially, cybersystems of the US.

From this point on, the process of institutionalization of cybersecurity as a policy issue became more evident in the US, absorbing other Internet governance issues, in the effort of the administration to *normalize* cyberspace. It started from the protection of critical infrastructures with the George W. Bush administration and it continued with the more normative and transparent approach, anchored to the White House, that led to the creation of new offices (like the cybersecurity coordinator and the new Cyber Command) with the Obama administration and the interventionist but erratic and often ineffective approach of the Trump administration. As it will be underlined, with a different degree, US presidential administrations have started to properly consider Internet governance as a global challenge with Obama. The Biden administration has the opportunity to consolidate that trend and counterbalance the structural weaknesses of the 'networked heterarchy' (Brousseau, Marzouki and Méadel 2012).

A more detailed reconstruction of the described process follows. After the September 11 attacks, the Bush administration included cybersecurity in its broader strategy for protecting critical infrastructures, and released its National Strategy to Secure Cyberspace in 2003. President Bush also signed the Homeland Presidential Directive 7 in 2003, establishing a national policy for Federal Departments and agencies to identify and prioritize the United States' critical infrastructure and key resources and to protect them from terrorist attacks. In 2004, the Congress allocated $4.7 billion to cybersecurity, thus achieving many of the goals stated in the President's National Strategy to Secure Cyberspace.

A Comprehensive National Cybersecurity Initiative (CNCI) was approved in 2008.[14] Its aim was to connect the cyberdefensive missions with law enforcement, intelligence, counterintelligence, and military capabilities to address the full spectrum of cyberthreats from remote network intrusions and insider operations to supply chain vulnerabilities. The CNCI described the government's offensive and defensive cyberspace goals and affected a host of federal agencies. With the cybersecurity Initiative, the CNCI indicated where the government's cybersecurity priorities were—on securing government and military networks from intrusion and disruption. It was crucial to the Obama administration, which used the directive as its starting point for cybersecurity from 2009 onwards.

In the United States, there is no centralized ministry in charge of science. Instead, many of its international science arrangements are made through the Department of State, where a leading role is played by a science advisor. This position has been central to efforts to engage the scientific community on issues important for the foreign policy community, including the negotiations for cybersecurity and the USA–EU Privacy Shield, a framework designed by the US Department of Commerce and the EU Commission to provide companies with a mechanism to comply with data protection requirements when transferring personal data from the EU to the US in support of transatlantic commerce. In the last fifteen years, program initiatives of the US presidents and secretaries of state have been based on S&T capabilities. The S&T goals set out in these documents are ambitious. What is missing, so far, is the commitment by the department to provide adequate resources. The need for smart policies is perceived also in the realm of economic policy towards California's high-tech industry as well as the national security policy, as an essential part of the science diplomacy activity.

Indeed, the president ordered an examination of the government's cybersecurity actions. The examination team summarized its suggestions in a cybersecurity policy review. To confirm cyberthreats, the cybersecurity policy review cited three cases: the failure of critical infrastructures (e.g., the disruption of electric power capabilities in multiple regions overseas, including a case resulting in a multicity power outage); the exploitation of global financial services (e.g., in 2008 a fraudulent transaction involving more than 130 automated teller machines in forty-nine cities within a thirty-minute time span); the systematic loss of US economic value (estimated in 1 trillion dollars from intellectual property data theft in 2008).

With the Obama administration, the US sought to reestablish its global normative leadership role through the area of science. In particular, science was expected to win back the support of foreign populations in Muslim countries (Hormats 2012). Though the chronological reconstruction of the evolution of Internet governance domain in the US continues to consolidate the relevance of cybersecurity, Obama started the elaboration of a global vision focused on the close relation between science diplomacy and Internet governance to tackle global challenges, underlying the potential of science diplomacy as a tool of soft power and to be 'a change agent.' (Van Langenhove 2016a).

The Obama administration distanced itself from its predecessor. It pledged to make its approach more transparent, affirmed that this would be anchored to the White House, and tried easing concerns about privacy (Sanger and

Markoff 2009; Vijayan 2010). Within the policy review on cybersecurity, Obama created a new office at the White House led by a cybersecurity coordinator (the 'czar' for media). This officer is responsible for organizing and integrating all cybersecurity policies for the government, working with the Office of Management and Budget (OMB) to ensure that agencies have funds for cybersecurity priorities, and coordinating the government's answers to the major cyberincidents.[15] During Obama's presidency, a new Cyber Command was established by the Department of Defense in April 2009. Its main task is to execute military cybersecurity operations and provide support to civil authorities. The defense Secretary at that time, Robert Gates, nominated the director of the National Security Agency as head of the new Command. In addition, the US Air Force established a cyberheadquarters and operational center with the power to launch cyberattacks when necessary and endowed with the mission of coordinating with other services cyberdefense.[16] Moreover, during the Obama administration a contrast between the Congress, trying to make information more transparent (after the Cybersecurity Act of 2012 for example) and the presidency more concerned with security, emerged.[17]

If the trend is to *normalize* cyberspace, the Trump administration slowed down the process, hindering the global vision of science and diplomacy as essential for supporting the universal scientific and democratic principles and for addressing global challenges. Moreover, the centrality of cybersecurity as the main asset of governmental policies dealing with cyberspace is confirmed, as it is shown by the consolidation of the institutionalization of cybersecurity also with the Trump administration. On the other hand, the strategy of interdependence is neglected by Trump, confirming, at certain degree, the idea of Farrell and Newman (2019) of the danger of the interdependence.

The Trump administration adopted a resolutely interventionist stance on cybersecurity policy, making it a tool of its foreign policy as well as for the internal US policy: 'I think that computers have complicated lives very greatly. I'm not sure you have the kind of security you need' (Diamond 2016). The presidential executive order 13800 signed by Trump on cybersecurity (11/05/2017) outlines the need to protect the Federal networks and direct all department and agency heads to coordinate their activities. Indeed, the perception of almost all the agencies of the US has been to consider themselves vulnerable to Cyberattacks. On 28 August 2017, eight NIAC advisors resigned en masse as protest against Trump's policy, considered inadequate to protect the digital security of the United States, as well as for the poor condemnation that Trump made of the Charlottesville incidents. Moreover, the attempt made by Trump of promoting a strong collaboration with Russia to create a cybersecurity unit 'to guard against election hacking and many other negative things' has been harshly criticized. Indeed, Trump decided to back away from the idea.

The Trump administration criticized the outgoing effort, strengthening strategic policymaking at the highest level. According to some observers, far from the 'American first' sort of isolationism, pledged by Trump during his electoral campaign, the tycoon president has adopted a resolutely interventionist stance, one even more aggressive than George W. Bush's (Hockenos, 2017). 'Promoting American values and countering authoritarianism in cyberspace' (Loyola and Glassman 2017) is therefore the next step in the ongoing battle for the control of the Internet. To win this battle, the efforts are aimed at reorganizing strategic apparatuses.

That approach of Trump's administration has been inherited by the new president, Joe Biden. A wind of change may come from the new administration, where President Biden deems to relaunch the role of the US as a beacon of hope in the international system, advancing the defense of the national interest as well as the international cooperation and a new multilateralism. Moreover, the Biden administration, in line with the conviction of the Madrid Declaration (S4D4C 2019), could relaunch the centrality of science diplomacy, applying on its renewed great potential to address global challenges. As mentioned before, the global reaction to pandemic crisis of Covid 2019 could have favored that evolution. At this stage, with a pragmatic approach, Biden is deciding whether to continue or deviate from the Trump administration's determination to implement cyber-offensive operations, in its efforts to modernize national cyberdefense.

The leading role of DHS's Cybersecurity and Infrastructure Security Agency in defending federal agencies outside of the Pentagon from hackers during the Trump administration has been confirmed, giving the agency a greater role in cyberdefense. On the other hand, the proposal of re establishing a cybersecurity coordinator position at the State Department, disbanded by Trump in 2018, confirms the determination of the Biden administration to state clearly that the US will respond quickly and proportionately to major hacks. At financial level, the strategic relevance of cybersecurity for Biden's administration is confirmed by the fiscal 2022 budget blueprint, including $2.1 billion for the Cybersecurity and Infrastructure Security Agency, as well as by the American Rescue Plan, signed by Biden in March, authorizing an additional $1.65 billion for cybersecurity efforts.

In May 2021, Biden signed an executive order to improve the nation's cybersecurity and protect federal government networks. This executive order makes a significant contribution toward modernizing cybersecurity defenses by protecting federal networks, improving information-sharing between the

US government and the private sector on cyber issues, and strengthening the United States' ability to respond to incidents when they occur[18].

However, the recent Colonial Pipeline incident, which temporarily shut down the pipeline supplying fuel to the eastern United States, is a reminder that federal action alone is not enough. Much of the domestic critical infrastructure is owned and operated by the private sector, and those private sector companies make their own determination regarding cybersecurity investments. Thus private sector companies have to follow the Federal government's lead and take ambitious measures to augment and align cybersecurity investments with the goal of minimizing future incidents.

THE EU STRATEGY: LOOKING AHEAD

According to Moedas, former European Union's Commissioner for Research, Science and Innovation, 'the EU approach to science diplomacy is simple. Investing in science and research for its ability to establish unity, as much as for its ability to stimulate intellectual and economic progress. Known in America for a long time, Science Diplomacy is an emerging term in the EU context, and a recent one at broader international level' (Moedas 2015). A year later, the European Commissioner contributed to the debate by writing an article on *Science Diplomacy in the European Union,* dedicated to the role of the European Commission as the executive body of the EU (Moedas 2016). At a time of great political uncertainty, a more political approach is necessary, defining a set of priorities, combined with a stronger political stance. In particular, Moedas devotes attention to the commission's *Europe as a global actor* in the international arena, where the complexity and far-reaching nature of highly politicized and culturally sensitive issues calls for a stronger scientific cooperation. In short, to achieve such goals, the EU must play 'an increasingly active and visible role in international science diplomacy' (Moedas 2016). Of course, Europe is not new to science diplomacy. In strategic policy areas—such as the nuclear—it dates back to the 1950s, when twelve nations signed on to foster the unifying power of science, both ideologically and pragmatically (Wedlin and Nedeva 2015).

Surely, efforts to implement ambitious program increased in the last decade, above all after the ratification of the *Treaty of Lisbon* in 2007, which refers formally to the creation of the *European Research Area.* Within a few years, we have witnessed a policy shift, with Europe's ambitions increasingly connected to the priorities of the international community: from the involvement in international climate negotiations to the cooperation in areas of public health, the European community is playing an important role in the global arena.

Is it enough? The report the *Tool for an EU Science Diplomacy* written by Luk Van Langenhove for the European Commission (2017) starts to answer the question. The overview of the concept and relevant tools and practices existing within the EU highlights that science diplomacy is *still scarcely known and perhaps not optimally used.* Findings suggest that 'most EU Member States do not have a Science Diplomacy strategy. In most cases, however, Member States are engaged in some activities that can labeled as Science Diplomacy (Witjes 2017). But also, the national efforts remain in most cases very limited and there are little support structures. On top of it, most national Science diplomacy activities are at best only loosely connected to Foreign Affairs policies. In other words, Science Diplomacy is not well developed within most of the EU Member States' as well as within the EU's own foreign policy and security policy. This is also due to the fact that strategic objectives of science diplomacy are based on national political interest while the EU is generally regarded as a 'perfectly suitable' global science diplomacy actor, it lacks an implementation agenda (Penca 2018).

It is noteworthy that the report offers useful recommendations to develop a proper EU science diplomacy strategy, *similar to what exists for EU cultural diplomacy*, focalized on three strategic areas, from which any reference to Internet governance domain is absent. In many ways, this omission is inexplicable, considering that the analysis is dedicated to the *Foreign and Security* policy as the main area to develop an EU science diplomacy strategy—that is, an area in which the EU has included cyberpolicy issues, developing documents, diplomatic initiatives and international agreements.[19]

Even if IG is absent in the principal documents where the attention of the EU has been focused, the EU adopted its own strategy on IG, as has been shown since 2005, thanks to the World Summit on Information Society. The European Commission articulated its vision on Internet governance in the document 'Europe's role in shaping the future of internet governance' (EC 2014). In it, the Commission promoted a number of principles, namely fundamental rights, democratic values and a single, unfragmented network. It also set key objectives, defending the multistakeholder model of governance, and strengthening a reformed Internet Governance Forum (IGF) and globalizing ICANN (Di Camillo and Miranda 2011). In 2016, the Commission proposed several legislative initiatives to 'Developing the Digital Single Market', including the reform of the copyright system. Since then, it has worked with social media companies on the creation of a Code of Conduct on on-line hate speech.[20]

Also, the first acts of cyberdiplomacy can be mentioned within the experience of the international debate on IG when the EU participate to it.

Progressively, the EU has broadened its policy scope to cover cybersecurity issues and to include them in its diplomacy, in a natural extension of its 'domestic' cyber-agenda to the international arena. The importance of cyber issues has been recognized in several EU policy documents and during the ENISA institutionalization in 2004. Debates and initiatives leading to the elaboration of fundamental documents, including the *Cybersecurity Strategy* (2013) and the *Council Conclusions on Cyber Diplomacy* (2015) have further highlighted this strategic relevance.

The Council emphasized a new priority and adopted specific conclusions on cyberdiplomacy. Cyberdiplomacy should be attained through an 'increased engagement and stronger relations with key international partners and organizations', including multilateral and regional organizations as well as third countries. The conclusions argue that the further development and implementation of a common and comprehensive EU approach for cyberdiplomacy at the global level is 'essential and crucial' (Council of the EU, 2015). Thus, the EU should add a strong cyberspace dimension to its external relations and its Common Foreign and Security Policy (CFSP). These conclusions articulate a vision for the EU's cyberdiplomacy based on the identification of five key priorities: the promotion and protection of human rights in cyberspace; the defining of norms of behavior and application of existing international law in the field of international security; Internet governance; enhancing competitiveness and prosperity as well as capacity building and development. A sixth priority refers less to the *objectives* of cyberdiplomacy, and more to its *channels* as it calls for a 'strategic engagement with key partners and international organizations' due to the 'global cross-cutting nature, scope and reach' of cyber issues (Council of the EU 2015).

Besides referring to IG, the document recalls the same principles and some objectives proposed by the European Commission in the mentioned document 'Europe's role in shaping the future of internet governance' (EC 2014). It highlights, therefore, the EU Commission commitment to building a strong relationship between science diplomacy and IG.

Similarly, former European Commissioner for Home Affairs Cecilia Malmström noted in 2012 that a fundamental objective of the EU's cyberdiplomacy is 'to reach out to its strategic partners to make our response more effective'. Therefore, when the 2013 Cybersecurity Strategy called for a 'renewed emphasis on dialogue with third countries' to establish deeper cyber partnerships, it was more a continuation than a revolution of the strategy implemented for the cyber domain, in line with the broader diplomatic efforts to focus on strategic partners. The Council of the EU (2015) urged the EU member states and the EU institutions 'to prepare cyber dialogues within the framework of effective policy coordination, avoiding duplication of efforts and taking into account the broader EU political and economic interests, collectively promoted by all EU actors', whereas the cyberstrategy insisted that 'EU consultations with international partners on cyber issues should be designed, coordinated and implemented to add value to existing bilateral dialogues between the EU's member states and third countries'.

The EU's strategic cyber partnerships with Japan and the United States are the best example of EU cyber bilateral dialogues. They have been developed on the basis of closer cooperation and the commitment to improve the bilateral structures and practices, including elements of cyber science diplomacy. In 2010, the EU and the US established a Working Group on Cyber-security and Cyber-crime; in 2014 the EU and the US, as well as the EU and Japan, held their first meetings of the EU-US and EU-Japan Cyber Dialogues, which have continued to meet since then once a year (Kadlecová, Meyer, and Ravine 2020). During the EU-US Cyber Dialogue in December 2016, the EU and US announced the promising creation of the Transatlantic Cyber Policy Research Initiative (TCPRI). However, as pinpointed by Kadlecová et al., 'both partners failed to deliver on their plans to take appropriate, timely action' (Kadlecová, Meyer, and Ravine 2020). In addition to this, the EU entertains programs to third countries and regions such as Latin America (Bonilla Aranzales 2017; Selleslaghs 2017), the Mediterranean region (Penca 2018) and others that can be regarded and studied as science diplomacy.

The EU's strategic cyber partnerships confirm our hypothesis of the radical shift in IG due to the massive rise of cybersecurity issues in the political agenda of national governments, as it shown by the EU engagement in new means and models of security policies, building relationships on cybersecurity issue with third states in international fora.

Cyberdefense, falling under the intergovernmental Common Security and Defense Policy (CSDP) mandate, is the newest and therefore the least developed initiative of the EU's current reorientation, that contributes to the international debate on IG. Following a soft power approach, Cyberdefense is based on the concept of security as resilience logic, linked to measures of self protections, access to military and civilian assets used for military purposes. Cyberdefense is not defined in the EU documents due to the different conceptions and sensitivity among member states on this issue. While in relation to cyberdefense EU–US logics are fundamentally different, on the issue of Network and Information Security (NIS)—and in particular critical infrastructure protection (CIP)—there is at least some sense of convergence on the need for regulation. The EU's approach to NIS has gradually moved from a hands-off meta governance to a voluntary approach to information sharing and reporting of major incidents thanks to numerous platforms. On the other hand, that convergence on data privacy has been considered 'uneasy' for the opposing attitudes, cultures, and legal systems of the EU and the US (Kerry 2014; see also Bygrave 2013). The mass surveillance security practices of the National Security Agency (NSA) were justified by a 'national security' logic and by US laws, supporting the US legalistic approach to cybersecurity with the US administration's approval. The failure of the 'Safe Harbor' agreement replaced by the Private Shield is yet another sign of this distrust.

The EU focuses on constructing a resilient ecosystem that enables not only protection through building capacity but also the ability to show flexibility and recover from cyberattacks. Moreover, the EU's system is characterized by the notion of security as resilience, where security does not mean the ability to build offensive capacity and defend the cyber perimeter, but rather adheres to the notions of creating adaptable, flexible, and robust systems and a complex regulatory environment with shared responsibility and multiple stakeholders (Coaffee and Fussey 2015).

Creating a culture of cybersecurity has been at the very center of the EU's efforts of achieving security in the information society (IS) for many years. The Budapest Convention provides a platform for a culture of cybersecurity to emerge on the legal and operational level based on a common, if not completely harmonized, set of minimum standards, definitions and protocols. The EU's approach has been essentially legalistic, focusing on cybercrime and with an emphasis on cyberdefense and soft power capabilities (Bendiek 2014; Christou 2016). Indeed, some have even gone as far as to describe the EU as a civilian cyberpower (Dunn Cavelty 2013).

The relevance and the depth of the EU's new reorientation in this field has started to consolidate for a number of reasons: an increased though still inadequate level of institutionalization, clearer rules of 'mutual recognition and mutual legitimization' that interconnect the presence of public and private organizations and the role of EU institutions in coordinating the actions of member states in the EU. The reorientation has been subject to a process of acceptation and acceleration. President Juncker created a specific Commissioner portfolio for the Security Union to ensure an effective implementation of the commitments made. Some relevant actions taken by the EU Commission in the last five years exemplify this process: the mentioned adoption of the Directive on Security of Network and Information Systems in July 2016, the adoption of the European Agenda on Security in September 2016, the proposal to revise the Schengen Information System (SIS) and the stepping up of the implementation of cybercrime acquis in December 2016, the Proposal to revise the Privacy Directive January 2017, the package of cybersecurity measures (the Cyberdiplomacy Toolbox) in September 2017, the Cyberdefense Policy Framework (2018); the EU Cybersecurity Act and the EU toolbox for

5G security (both 2019); as well as the EU Security Union Strategy and the Screening of (Digital) Investment (2020).

As outlined by Kadlecová, Meyer, and Ravine (2020) the Cyberdiplomacy Toolbox 'introduced new initiatives to further develop European cyber response and resilience — among others, strengthening the role and mandate of the EU Agency for Network and Information Security (ENISA), introducing a cybersecurity certification scheme recognized across the EU Member States, and prompt implementation of the Directive on Security of Network and Information Systems (the NIS Directive)'. On a more general scale, the EU Cyberdiplomacy Toolbox is important in its declared goal of promoting the application of international law in cyberspace, responsible state behavior, and greater cooperation and more agile joint EU diplomatic reaction to malicious cyberevents. The package has designed its own countermeasures and sanctions regime against IT attacks to respond to cyberattacks as it has been the case in July 2020 in the course of the handling of the 2015 hacker attacks on the German parliament (Kadlecová, Meyer, and Ravine 2020).

The European Commission's proposal for a European Cybersecurity Competence Network and Centre also supports some of the current projects. The main purpose of this new initiative, which is funded under the next multi-annual financial framework for 2021 to 2027, is to 'help the EU retain and develop the cybersecurity technological and industrial capacities necessary to secure its Digital Single Market' while increasing 'the competitiveness of the EU's cybersecurity industry and turn[ing] cybersecurity into a competitive advantage of other European industries' (Bendiek and Kettemann 2021).

Finally, the 16 December 2020, the EU updated its 2013 cybersecurity strategy, extending it to a strategy for cyberspace that should make *the norm of state responsibility a cornerstone of its approach*. With the evolution of the last years the EU institutional architectures, the current division of powers is starting to change, overcoming one of the most significant structural limit, because a stronger EU at the political, institutional, and normative level is arguably the only possibility that the EU has to remain *a common space of freedom, security and justice* as stated in article 1 of the Treaty of Lisbon, defending and promoting EU values against the oligopolistic and private interest of big global digital firms, like Google, Microsoft and Facebook.

The last evolutions described (strengthening the role and mandate of the ENISA, introducing a cybersecurity certification scheme recognized across the EU member states, prompting implementation of the NIS Directive, as well as the focus on fundamental rights, democratic values and resilience, the quest for applying international law in cyberspace and the proposal for a

European Cybersecurity Competence Network and Centre) have confirmed the trend toward the emergence of the notion of Internet governance as a global challenge.

Notwithstanding the importance of a diplomatic service for the European Union and the relevance that cybersecurity has gained in the European policymaking, there are still some ongoing weaknesses (Council of the EU, 2015). First, scholars outline that the most important foreign policy invention in Europe, the creation of a High Representative for Foreign and Security policy and an External Action Service (EEAS), brings together—for the first time—national diplomats, civil servants from the Commission, and officials from the Council secretariat under the same roof.[21] However, 'it is comparable in size to the foreign service of a medium-size member state', and often in contrast to national diplomacy (Adler-Nissen 2014). Regarding the cybersecurity issue, the entry of the EU in the cybersecurity domain caused a visible institutional change in the operational and regulatory status of the European networks, but the institution-building activities have halted or slowed down (Ruohonen, Hyrynsalmi, and Leppänen 2016). Thirdly, 'to protect critical infrastructures, existing EU law and the 2016 EU Network and Information Security Directive (NIS Directive) are to be revised, and greater use will be made of artificial intelligence to identify cyberattacks against hospitals, utilities, and transport networks. In order to achieve the "strategic openness" and the protection of the digital single market, the EU cyberdiplomacy should be made more coherent in its supranational, democratic, and economic/technological dimensions' (Bendiek and Kettemann 2021).

CONCLUSIONS

This chapter started to address the critical topic of the intersection between science diplomacy and Internet governance, with the centrality of the institutionalization of cybersecurity. We may wonder whether Josef Nye's comment (Nye 2014) on the lack of general consensus on global cybergovernance is still working. The reconstruction of the evolution of Internet governance domain in the US and the EU in the last years has verified the rising relevance of cybersecurity policy, reformulated and widened in its scope to include different policies. The hypothesis of the radical shift due to the massive importance of CS on the political agenda of all the actors involved has been confirmed. On a more general note, however, the low level of the institutionalization of IG, its dynamic nature, and the rising role of states in cyberspace have been hindering the nature of Internet governance as global challenge.

The state of transatlantic collaboration in the area of cybersecurity strategy is mixed. On one hand, it is evident that there are still divergences between the US and the EU, rooted in political, social, and cultural differences and in some case almost opposing values promoted by the respective legislations. Ülgen (2016) outlines this divergence on a wide spectrum of public policies, from national security to privacy, from data confidentiality to data transfers, from taxation of big Internet companies to cyberwarfare norms. They have adopted their own global cyberstrategies but without a serious engagement for collective actions.

On the other hands, the US and the EU can take the lead of western democracies by setting up a multilateral regime to limit and to oppose illiberal regimes, using cybersecurity resources. However, this is not simply a case of the 'West' vs. the 'Rest'. It is apparent, in different spaces, and it is constantly shifting between varying coalitions of actors. Thus, whilst the EU and the US certainly agree on the need for an open, accessible, and resilient Internet and a multistakeholder model for Internet governance, in contrast to states such as China and Russia that promote a state-driven, intergovernmental system, there are clearly also disagreements which materialize from the security logics that underpin their approaches to cybersecurity.

Indeed, the US approach to cybersecurity has at its very core the concepts of military defense and deterrence which includes the dimension of offence (Lewis 2016), whereas the EU approach is strictly linked to the dimension of defense. Against this backdrop, the revitalization of bilateral cyberdiplomacy between the EU and the United States has gained special attention since Joe Biden's election as US president. From the new US perspective, the chances for the EU to defend the internal market in the digital domain against China and other authoritarian regimes rests on its willingness to establish an alliance of democratic multilateralists committed to a free, open and democratic internet, 'The digital version of the Schengen Agreement', already proposed by Clarke and Knake (2011) with the US-led 'Internet Freedom League', that must include the United States, Canada, Australia, Japan, and others, even if they only cooperate in the short term with ad hoc coalitions (Bendiek and Kettemann 2021).

In the United States, notwithstanding the evolution and the empowerment of science diplomacy, also in the sphere of Internet governance, the focus is mainly on the security aspect of the problem. The US cybersecurity system, centered on the returning to state power of the normalization process, aiming to be considered a model for the other actors of the international community (starting from the EU), needs to overcome some problematic features still unsolved.

On the other shore of the Ocean, in a process of normalization that implies the explicit action of power politics of states, it must be said that the EU could risk of being caught in a weak position, due the structural fragility of the Common Foreign and Security Policy, where cybersecurity should be dealt in an EU perspective. It has been already underlined that the science diplomacy in the cyber field remains weakly institutionalized: the goal of 'strategic autonomy' of cyber science diplomacy might hinder international cooperation with third countries, science diplomacy may fail to mobilize other EU official arenas still not engaged in the field of cybersecurity, and it has registered a consistent turnover among EU officials' working in various departments of the Commission and the EEAS (Kadlecová, Meyer, and Ravine 2020).

However, some initiatives have been implemented to address key cyber-policy challenges and increase policy research capacity on cyber issues (Kadlecová, Meyer, and Ravine 2020)[22]. What has emerged in the EU is a promising hybrid governance attitude across and within the main priority areas of its cybersecurity strategy (Christou 2016). The significant barriers existing among states in terms of culture, institutional capacities, resources, legal frameworks, as well as a poor understanding and awareness of the potential cyberthreat (ENISA 2014) are being addressed with a large use of resources in the last years and a significant process of institutionalization.

The EU is starting to recall to its federalist perspectives, in line with the tradition of the founding fathers of the EU who overcame difficulties of the European unification by deepening the integration process with more power to the common institutions. Thus, this prospect would reinvigorate an ancient ambition of the federalist movement. Today, the same spirit that animated leaders like De Gasperi to enthusiastically accept the European defense Community (EDC), could regain momentum and lead to a European system of defense and Security with a cession of power from member states (though the EDC failed to obtain the ratification of the French Parliament, it represented a great step towards a European defense system).

Changes generated by the pandemic emergency seem to favor a change of course both in terms of politics and policies. Just as they seem to favor it, relaunching the transatlantic partnership, the provisions and guidelines of the Biden administration. These are certainly important news, but it is still too early to say how they will impact, even in the short term, on the redefinition of the overall architecture of powers and, consequently, on how Internet governance and cybersecurity policies will come into effect as part of European science diplomacy.

NOTES

1. Many policies of governments can be considered from the perspective of science diplomacy.

2. The internet is mentioned only once, as a tool, with other social media, to increase the influence of the new actors and networks into the blogosphere and in international arena.

3. Brousseau and Marzouki (2012) mention a 1998 ITU resolution as the first reference to Internet governance. Hofmann, Katzenback, and Gollatz wrote (2014) that 'the use of the term Internet governance in academia started around 1996 in the US'.

4. A few years ago, Mueller, Schmidt, and Kuerbis (2013) raised doubts about the 'claim that introducing security concerns into Internet governance necessarily leads to more hierarchy and/or a greater role for national governments'. Recently, Dunn Cavelty (2015), instead, have observed that: 'Overall, cyberspace has been upgraded to a strategic domain whose development is no longer left to non-state actors'.

5. For an up-to-date analysis of the OECD evidence, see the report by the Diplo Foundation (2016), where it is possible to appreciate the infographic expressing the results.

6. The transformation of core executives would be a good starting point of the analysis in comparative perspective, considering that the main institutional and organizational transformation occur at this level. A first recognition of the scientific literature underline the central role played in the US by the President and China by the Central Government, whereas in the EU there is a fragmentation of responsibility among member states, who have the real power, and the EU institutions promoting harmonization and coordination of state policies.

7. Indeed, in the last decade a relevant evolution at the core this intersection, strategic for the Science and Diplomacy relationship, can be found in the examples of Big Data and artificial intelligence, which have already become a consolidate reality to deal with. They can be conceived as a tool for diplomacy, a new topic on the diplomatic agenda and the ways in which data changes the geopolitical environment in which diplomacy is situated, expanding in areas like cross-border privacy, e-commerce, and international cybersecurity (Rosen Jacobson, Hone, and Kurbalija 2018).

8. For a critical analysis, see (Jeutner 2019).

9. The GFCE assesses different initiatives in four domain: e-governance, data protection, cybercrime and cybersecurity. For the e-governance and data protection there are not GFCE initiatives. Regarding the cybercrime, the relevant initiatives are only the *Cybersecurity and Cybercrime Trends in Africa* and *Preventing and Combating Cybercrime in Southeast Asia*. On the other hand, for the cybersecurity the considerable number of eleven initiatives have been started: (1) Assessing and developing cybercapability; (2) CSIRT Maturity Initiative; (3) Critical Information Infrastructure Protection Initiative; (4) Cybersecurity and Cybercrime Trends in Africa; (5) Cybersecurity Initiative in OAS member states; 6) Global Campaign to Raise Cybersecurity Awareness; (7) Critical Information Infrastructure Protection Initiative; (8) Internet Infrastructure & Maintenance Initiative; (9) Progressing Cybersecurity in Senegal

and West Africa; (10) Promoting Cybersecurity Due Diligence across Africa; (11) Responsible Disclosure Initiative (Ethical Hacking).

10. Indeed, the first experiences, since WSIS in 2003, fostered a vision of the relationship between IG and diplomatic action strictly linked with the political and cultural orientation dominant at that time, encouraging such expectations. See (Kurbalija 2008; 2016).

11. In September 2015, the US and China concluded an important agreement facilitating cooperation on cybercrime issues and aiming to curb cyber-enabled economic espionage. See Maurer (2016).

12. The EU and the US have traditionally had similar normative positions on the Internet and its governance. They both adhere to the value of a global Internet as a public collective good that should be available to and accessible by all citizens (Christou 2016). The 2015 US-EU fact sheet on cyber cooperation aimed at governing cyberspace highlights the extent to which 'cooperation is founded on shared values, interest in an open and interoperable Internet, and commitment to multistakeholder Internet governance, Internet freedom, and protecting human rights in cyberspace'. However, Edward Snowden's release of classified US documents in the summer of 2013 stressed the solidarity between the EU and US and the transatlantic cyber collaboration.

13. While the department has taken important steps in strengthening S&T capabilities, in Washington emerges a 'tale of two states', considering the activities of the US embassies abroad. They result being very slow in implementing S&T as a key factor of their diplomatic action. Numerous US national reports have underlined the decreasing dominance of US STI, asking for the rising of domestic STI budget, but this is highly unlikely to happen any soon due national financial constraints. If the United States can no longer be assured of leadership in STI, all alternative ways of maintaining its leadership are *through synergistic partnerships.* (Colglazier and Lyons 2014).

14. Authorized through the National Security Presidential Directive 54.

15. The Coordinator unifies the Army, Air Force, Navy, and Marines. The Obama administration subsequently appointed Howard Schmidt as the White House cybersecurity coordinator, and, in 2011, the US Defense Department developed its own cyber strategy.

16. Nonetheless, some observers have argued that the strategy is insufficient because it lacks a unified approach and specific funding sources (Nakashima 2011). In his recent book on cyberwar, Richard Clarke, former US counterterrorism czar, concludes that the international community should develop cooperative strategies for dealing with the new state of international cyber affairs (Clarke and Knacke, 2011).

17. After the scandals of NSA massive violations of privacy and civil liberties, Obama's public statements in defense of the human rights, reaffirming at the same time the commitment to net neutrality to keep internet open and free, have been considered inconsistent.

18. Specifically, with the executive order the President is signing Modernize and Implement Stronger Cybersecurity Standards in the Federal Government, Improve Software Supply Chain Security, Establish a Cybersecurity Safety Review Board,

Create a Standard Playbook for Responding to Cyber Incidents, improves the ability to detect malicious cyber activity on federal networks, Improve Investigative and Remediation Capabilities

19. It is significant that the Commission has paid increasing attention to science diplomacy as part of its foreign policy, although its driving force is the Directorate General for Research, not the European External Action Services.

20. The European Parliament, the Council and the Commission have been standing for an inclusive approach, safeguarding the multistakeholder model, based on the full involvement of all relevant actors and organizations, even in a period of large-scale Internet surveillance and cyberspace difficult relations (EC 2014). The second of the *Internet Governance Strategy 2016-2019* proposed by the Council is entitled '"A continuum of core values": democracy, human rights, and rule of law in the digital world. Who among us does not agree?', 2016 closed with the first Internet Governance Forum (IGF) in Guadalajara with a large delegation of European Parliament, the entry into force of the first ever EU rules on net neutrality, protecting the right of every European to access internet content without discrimination and the adoption of the Network and Information Security Directive, the first EU legislation on cybersecurity.

21. Some progresses have been registered in recent years, with the development of the European External Action Service (EEAS) department specialized in cyber issues, in charge of advocacy at NATO and the OSCE and that provides support to member states that have not developed their own capacities in the field (Bendiek and Kettemann 2021).

22. Promising initiatives in EU-US cyber relations that gives hints of the development of cyber science diplomacy are the EU-US Dialogue for Research and Innovation in Cybersecurity & Privacy (AEGIS), that was funded under Horizon 2020 and begun in 2017, and the Transatlantic Cyber Policy Research Initiative (TCPRI), that, launched on the 2016, has been reconsidered and discussed in December 2018 to device a new model for implementation

III

CASE STUDIES OF INTERNET GOVERNANCE DIPLOMACY

Chapter Eight

Free Trade Governance and Data Flow

The TiSA's Agreement Negotiation as a Case for Governmentality Studies

Maria Francesca De Tullio and
Giuseppe Micciarelli[1]

The present work explores the outer edge of Internet governance where this seems to overlap with trade law and, in particular, with a latest generation of Free Trade Agreements (FTAs). To be more precise, the research investigates whether and how the latter could contaminate the former. FTAs provide a solid structure for diplomatic public-private relations, whose key articulations could be assumed in practice as a successful example for digital issues.

AN INTERDISCIPLINARY METHODOLOGY
FOR INTERLACING FIELDS

FTAs are a vast and multifaceted phenomenon, which cannot be reduced to a single matrix. However, this study only focuses on a species of that category, i.e. the latest wave of FTAs, which includes, for example, the *Comprehensive Economic and Trade Agreement* (CETA) and the *Transatlantic Trade and Investment Partnership* (TTIP). That model attains the most advanced institutionalization of the international state-business cooperation (Micciarelli 2017, 252–54).

The research raises many unanswered questions: how do the institutions created by these Treaties relate to the traditional democratic and representative ones? How do private subjects participate in their deliberations? What are the new diplomacy factors, forms of lobbying, and relations of power?

The study does not give any definite answers to the issues raised, but thematizes them within a specific theoretical framework, namely the 'governmentality' developed by Foucault. This category can be used to discern the institutional transformations brought about by the rules of trade, which could reflect on Internet governance and democracy in general.

To that aim, the analysis adopts a method based on law and on legal and political philosophy. It proceeds through a survey of literature and documentary materials, including legal acts, jurisprudence, and official texts. Given that the negotiation process of FTAs is classified, their content is largely uncertain. However, drafts have been partly leaked by civil society organizations (e.g., WikiLeaks and Greenpeace) and some parties' official positions have been voluntarily released.

Given these premises, the work is organized around a case study. The research focuses on the negotiations over the data flows in the *Trade in Services Agreement* (TiSA).

The TiSA is a FTA that is still being negotiated, it seeks to minimize non-tariff barriers to international trade in services. Although the agreement has yet to be approved, we can hypothesize that it would adopt the model of the new generation of treaties as these have already attained the most solid practical results.

We have chosen this agreement as a test because, due to its broad geographical and material scope, it is likely to have a huge impact. On one hand, the TiSA would involve twenty-three Parties, accounting for about 70 percent of international trade in services.[2] It is also an open Treaty and is susceptible to being integrated in the World Trade Organization (WTO) framework, in case a critical number of WTO Members would adhere[3]. On the other hand, the material scope of the Agreement covers many key web-related matters: not only the transnational data flow, but also, for example, the localization of computer facilities, the discipline of source codes and net neutrality. Among these subjects, the data traffic receives particular attention in our study, as it best shows how trade law intersects a plurality of personal and political freedoms.

Data flow is an increasingly important issue in international relations. Indeed, the applicable law (e.g., privacy and data surveillance regulations) depends mostly on the localization of the information. This is not a trivial question because a far-reaching control on data entails a greater power for States and private entities. For example, it could allow advertisers to track customers' preferences, insurers to assess risks, citizens to monitor the administration's actions, governments to keep the population under surveillance, and so on. Moreover, as more business models and practices move onto the digital platform and data becomes increasingly shared and exchanged on an international scale, its relationship to international trade intensifies. Since data is gathered, digitized, stored, and moved on a truly global basis by a multitude of parties, restrictions, and regulations concerning data directly affect global trade (UNCTAD 2016, 3–4; *See* also Singh, Abdel-Latif, and Lee Tuthill 2016, 2).

Therefore, as explained in paragraph 3, information is a very peculiar commodity, because it affects economy and human rights hugely on a global scale.

Having stated the reasons underlying the choice of our case study, an outline of the chapter's structure is in order.

For clarity's sake, the case study is introduced in paragraph 2, which sketches a panorama of the trade agreement system, pinpointing the main issues at stake.

Against that general background, paragraphs 3 to 5 conduct a more in-depth observation of the case study. Paragraph 3 in particular illustrates the political stakes of the Treaty in the data field. This is only an example to make a more general point: trade law and Internet governance, while affecting technology regulations, also impact fundamental rights. Afterwards, paragraph 4 shows how the TiSA is likely to recast the balance between these fundamental rights and economic freedoms. In that regard, the draft Agreement is found to state a rule-exception paradigm that is radically different from most democratic Constitutions. Finally, paragraph 5 analyses the institutional side of the TiSA and connects it to some critical transformations that FTAs are bringing to the constitutional states of right.

The conclusions of this work (paragraph 6) seek to frame the case study in a broader theoretical background. Namely, they employ the heritage of the 'governmentality studies' to approach the FTAs phenomenon and its intersections with Internet governance.

THE EVOLUTION OF NEOLIBERAL INSTITUTIONAL DESIGN

Globalization has ineluctably generated an unprecedented degree of global interdependence among subjects and systems: organizations, private and public actors, institutions, and networks within different functional systems. These intersections have generated new institutional arrangements with a vocation to follow and develop new spatial and temporal trajectories for decision-making. This process is political, economic, cultural, and technological at the same time; it is multi-causal and produced by distinct yet interconnected phenomena that have occurred at different speeds, in different sequences and in different places (Aglietta 1998).

In particular, globalization has taken its actual appearance from the intertwined revolutions in two fields: technological and economic. These two realms are closely linked because the evolution of the first involves a mutation of capitalism (Deleuze 1992, 6). The technological field is never-ending and escapes any attempt of central governing. Here technology plays an important role because it affects the way 'behaviors are directed, constrained and framed by resources such as algorithms, content management systems and operating systems' (Badouard, Mabi and Sire 2016, 9). Regulatory problems are crucial for any interdependence system, and this is evident in both sectors of technology and economy and is further increased by the dimension and complexity of the actors involved *and the unpredictability of innovations*. The birth of the internet is probably the paradigm of this landmark transformation. Referring to the economic field, exchanges of goods, capitals, and services spread across a *global market* has become a synonym for *globalization*. Symmetry and confusion between these two distinct concepts once more show how capitalism has seamlessly occupied the horizons of the thinkable (Fisher 2009). The interdependence of economies does not mean that they are symmetrical at all, rather the opposite. Neoliberal rationality is not only economic but also political (Dardot and Laval 2017; Bazzicalupo 2015); and for this reason it was able to promote the building of a global trade institutional regime that has almost linked the entire planet in *a special kind of cooperation for competition*. If it is true that we live in an era of interdependence (Keohane and Nye 2011, 3), neoliberalism has overlapped this concept with a global system of international relations built on the free market, thus conveying a distorted image of interdependence[4]. This is one of the main issues with the neoliberal approach to interdependence. The context of global commerce has generated models of governance that have a potential paradigmatic nature, becoming the global political laboratory for the modeling and transformation of new institutions that have reshaped the system of international relations. Our hypothesis is that the last generation of Free Trade Agreements—that are presented as the most developed forms of supranational institutionalization as a rule-making laboratory (Sauvé 2014, 15)—have given rise to a new evolution of governance that represents the apex of neoliberalism and the turning point of its crises; this is quite evident if we look at the standstill of TiSA and TTIP during Trump's mandate[5]. As anticipated, we refer to trade regimes as a heuristic simplification to name a larger number of agreements and para-institutional forms of cooperation, differently articulated, covering the exchange of goods, services, and capitals: a pluriverse of 'institutions of globalization' (Ferrarese 2000) that overlaps and even exceeds the frameworks of the WTO, World Bank (WB) and International Monetary Fund (IMF). We can think, for example, of the nearly inextricable texture of Bilateral Investment Treaties (BITs) and Multilateral Investment Treaties (MITs) which have forked in *institutional chains of global interdependency* and have revolutionized one of the fields that is among the most conflictual and sensitive to the category of national sovereignty, not only in the Third World (Anghie 2004) but also of the richer nations today.

The regional and transnational plurilateral agreements are even more relevant in our argumentation for the institution of free exchange areas outside the WTO framework. A common element among them is the remarkable degree of creativity and institutional experimentation: cooperation bodies, stakeholders' committees, and a panoply of arbitration court procedures for conflict resolution are a constant necessity in order to create a common framework within a cooperation system. Among the many legal changes brought into these Free Trade Agreements, one of the most significant is the creation of new political bodies in order to fulfill both the legal harmonization and the demolition of the so called 'non-tariff barriers to trade', that is to say, the constellation of norms and standards that rule the production and export of goods and services according to different national-based legal systems. In this direction, they represent an ideal institutional design, because they prove apt in regulating a 'global order which is no longer an *inter-nations* order, but a complex scheme of interaction among mixed actors, where traditional subjects are put in relation with new ones, equally or more powerful, which represent different interests' (Catania 2008, 165).

Generally speaking, alternative plurilateral models include at least five types, whose differential criteria lie in the degree of institutionalization and the relation with the WTO normative system (Peng 2013, 5–7). Here, we only refer to the ones belonging to the latest generation: the CETA, the TTIP, and the TPP[6], that we can address as the core of 'free trade governance' model (Micciarelli 2017, 261). Indeed, these kinds of agreements are the most ambitious, not only for their extension, but also for the system of permanent neo-institutionalization that characterizes them, by virtue of which they are referred to as 'living agreements'. Indeed, it has been observed that 'CETA and—probably—TTIP are going to be very different from recent FTAs in terms of regulatory coherence and cooperation' (Chase and Pelkmans 2015, 25). These ones, moreover, 'if used to create a standing transatlantic regulatory laboratory, might indeed emerge as a model of IRC that may spread beyond TTIP toward the creation of a global policy laboratory' (Wiener and Alemanno 2015, 133).

Free trade governance is one of the few and concrete forms of institutional multilateralism, and this gives them an advantage over other kinds of supranational institutions that reveal themselves to be utopian and, as a matter of fact, controversial projections of a global democracy which is hardly observable in practice. In politics, praxis is often stronger than theory and 'among major global regimes, the trade regime has been the most successful at achieving its announced objectives (liberalization and nondiscrimination), while continuing to widen its scope—from tariffs to non-tariff barriers, from goods to services. It has also become the most legalized of the major international economic regimes, with the most elaborate formal procedures' (Keohane and Nye 2001, 3).

This trend is still strong, despite the crisis of neoliberalism and the grow- ingly influential condemnation of its evident inequalities and responsibility for climate change and environmental disruption (Stiglitz 2002; Chossudovsky 2003; Chomsky 2016). Despite these criticisms, the neoliberal architecture of trade agreements can represent a model for other sectors in pursuit of multi- stakeholder systems of governance. It has been recently observed that for its part, the Internet governance community is realizing that trade negotiations are not only about goods and services but are also moving toward governing deeper regulatory issues extending beyond national borders, which include intellectual property, data protection, privacy, and cross-border data flows (Singh, Abdel-Latif and Tuthill 2016).

THE POLITICAL STAKE OF INTERNET GOVERNANCE: THE DATA FLOW ISSUE IN THE TRADE IN SERVICES AGREEMENT

The above-described regimes are examined here as potential source of inspi- ration for the Internet rules. A basic assumption of such an investigation is that FTAs, not differently from Internet governance, are not merely sets of techno-economic arrangements. Rather, they entail political effects, affecting the juridical values enshrined in national and supranational legal systems.

More generally, we deem that no Internet governance is neutral, since 'techniques are always instruments and weapons, and because of the very fact that they serve everyone, they are not neutral' (Schmitt 1932, 178–79). Indeed, effectiveness is the objective of every technical arrangement, but the measure of that effectiveness can only be given by the purposes (De Minico 2012, 293) and the social imaginaries (Mansell 2016, 22) that each technology is intended to enact. Then, each use of ICT implies a previous hierarchization of the concerned interests, not only in the shaping of the infra- structure, but also in the drafting of the relevant rules.

As to FTAs, their leading objective is to lower non-tariff barriers. Our hypothesis is that this gives a peculiar political direction to how they disci- pline the web. Specifically, this imprint pushes to prioritize free market over other instances, disrupting the traditional democratic balance.

An exhaustive discussion of this theme would exceed the scope of this work. Yet, as explained in the introduction, we are employing a case study to depict some evidence of the claim. In particular, we are assessing how the TiSA, now in course of negotiation, could impact the regulation of data flows.

Data is emblematic here, because its technical substance does not agree with its concrete economic and legal destiny. More precisely, information is deemed a public good by those who focus on its nature. Still, it circulates as a private good, subject to legal 'enclosures' and appropriation processes (Newman 2014, 411–40).

The economic literature analyzing the raw substance of data approximates it to a non-rival and non-excludable good (Śledziewska and Wloch 2017, 265–68): it can be duplicated and transferred virtually everywhere at almost zero cost and is not consumed by utilization. Furthermore, the innovativeness of modern data analysis instruments derives exactly from that possibility of reusing the data sets. Indeed, 'data mining' does not need specific samples suitable for each experiment but allows us to scrutinize huge quantities of data collected from different sources and with different aims, even in absence of a working hypothesis prior to the creation of the data sets (Mayer-Schönberger and Cukier 2013, 11–70 and 147–58). Then, not only is information non-rival, but it also delivers its highest potential value if shared, reused, and linked with other databases.

A second characteristic of data is that it can serve multiple purposes that meet different and sometimes contrasting interests, involving either the economic, personal, or public sphere. For example, data showing netizens' political behaviour can serve many functions. On one hand, it can be stored by the platform owner for business purposes, i.e. to match advertisers' need for profiling and targeting potential buyers. On the other hand, it can be used to perfect opinion polls and so improve the quality of information.

That said, we can conclude that data is naturally prone to circulation and sharing and that it affects plural vested interests. Now, if non-tariff barriers rise in this field, it is because different legal systems balance the juridical values involved in different ways. These are the very norms that TiSA needs to harmonize. To do that, the Agreement needs to settle *its own regulatory balance*, which cannot but be a political decision.

After all, trade law already affects digital rights. In addition, according to some authors, it is by now the most successful attempt to govern contested Internet issues through international legal regimes (Goldsmith and Wu 2006, 167). Ultimately then, the FTAs deal with online freedoms and Internet democracy itself, which 'does not follow automatically from the values associated with new technical arrangements for online information production and sharing' (Mansell 2012, 60).

To gain a better insight of the claim, a brief account has to be made of the concerned interests.

A first set of positions to be conciliated includes different visions of the data economy. On one hand, some businesses, especially in the Silicon Valley, need data to be surrounded by 'enclosures'—namely exclusive intellectual property rights (De Filippi and Bourcier 2013, 3–6)—in favour of the people who collect them. Indeed, they profit from providing free digital platforms in exchange for users' data and then selling the gathered information for advertising purposes. On the other side, a growing number of EU stakeholders argue that data should be shared within industry, in order to innervate the whole agricultural and manufacturing sectors and deliver the maximum of wealth[7]. So, even if both positions advocate the 'free flow of data', the underlying visions remain divergent. The former intends this objective as a safeguard for commodification and appropriacy of information. The latter, instead, conceives it as a competition regime that proactively promotes the sharing of business data, through multiple access rights and data portability obligations (Drexl 2016, 30–38).

In doing so, the present situation shows that the creation of exclusive rights over data leads to concentrated markets[8]. This is a consequence of indirect network effects, that make the advertising revenues flow towards the already dominant actors. Indeed, on one hand, advertisers prefer to invest in the most frequently used platforms, since they acquire more data, thus guaranteeing a more effective targeted marketing. On the other hand, the incumbent's service is more appealing to consumers, because the dominant player hosts a larger community and has more profits to invest in quality. Consequently, a 'snowball effect' hits the newcomers, who fail to gain customers and are consequently unable to gather data in order to procure advertising venues and cover the initial costs (Orefice 2016, 718–33; Autorité de la Concurrence and Bundeskartellamt 2016, 17–51).

Concentration and distribution in data markets are in turn intertwined with basic liberties, which are the second set of positions affected by the regulation of the digital information flow.

The data subjects' privacy is probably the most directly involved. Individuals have an interest in preventing the nonconsensual circulation of their personal information. Dominance in the data market also gives more power for the businesses to impose harsher privacy terms to the users (Newman 2014, 442–43). But equality is also questioned by data oligopolies, if we consider that informational asymmetries enable private powers to orient other people's behavior on large scale. Not to mention the effects on popular sovereignty, due to secret surveillance. Indeed, in many juridical systems, data gathered by Internet operators are repositories that can be compulsorily and secretly accessed by intelligence agencies. Thus, the regulation of data flow incidentally tips the balance in the conflict between security, on one hand,

and privacy and government accountability, on the other. Moreover, data also affects political and electoral campaigns, since it is used to provide targeted information that can foster echo chambers (Visco Comandini 2018, 12–13; Claussen 2018, 133; Morozov 2017) or distort political accountability. For example, the Cambridge Analytica scandal revealed that politicians are able to target their constituents in an individualized way through social networks (Grassegger and Krogerus 2017).

Finally, data economy even changes labor law, with the creation of new forms of free or low-paid 'digital labor'. They can be simple tasks such as posting or 'liking' contents on social networks, or more complex ones on gig and sharing economy platforms. In these cases, through the collection and sale of users' data, time spent by consumers in online activities is quantified and turned into a source of profit by the platform providers, without any remuneration for the 'laborers' (Collin and Colin 2013, 35–57). Indeed, 'intersubjective ties are turned into a source of value, too' (Casilli 2015, 32).

This further strengthens the initial hypothesis of this paragraph because it shows that the horizontality of the technological architecture is not decisive. Even sharing activities can generate inequalities, due to the regulation of economic processes (Vercellone et al. 2015, 61–110). One of the reasons is that the design of ICT inherits all the ambiguities of the 'social imaginary' that animated the Internet of origins: while great emphasis was given to the collaborative creation, less attention was paid to the either exclusive or collective appropriation of the final products (Cardon 2010, 32–33; Mansell 2016, 22–23).

Hence, conclusively, every rule for the online economy covers many twisting and opposing rights. So, FTAs have in fact a political relevance, even questioning the democratic identity of some juridical systems.

TISA AND DATA FLOW: PRIVACY AS EXCEPTION TO FREE TRADE

We can now examine through the same case study what kind of impact FTAs exert on the juridical and political interests involved. The analysis mostly relies on the drafts of the TiSA, leaked in 2015 and 2016. Our conclusions take into account that the outcome is still aleatory, because negotiations are secret and still in progress.

The agreement is crafted to use a straightforward mechanism of rule-exception to address all the conflicts between economic and fundamental rights. Namely, free trade is stated as a straightforward general principle, while basic freedoms, where ensured, are provided as an authorization for the parties to derogate the treaty. This is not secondary, because, according to juridical science, exceptional rules have a narrow application, that shall be punctually justified and cannot be interpreted extensively or analogically.[9] On the contrary, general rules can only be set aside when they are expressly excluded.

In this sense, TiSA would be a fully original juridical order. Firstly, it would be radically different from many democratic Constitutions. Within these rights, personality rights and not free trade are 'fundamental' and so enjoy an 'assumption of maximum extension' (Grossi 1972, 15–16; Caretti 2005, 104). This means freedoms are applied in all occasions, unless an exception is clearly provided by the law and is apt to pursue an equally fundamental principle, and results are necessary and proportional to the benefit provided[10]. Secondly, the TiSA diverges from other pieces of international regulation on data protection, such as the OECD Privacy Principles. Therein, the pivotal aim is to impose a plafond of minimal safeguards for liberties, even if their *occasio legis* is to foster trust as a precondition to free exchange. On the contrary, the TiSA is a negotiating field where some Parties seeking to access new markets accept limitations of their demand to safeguard individual rights (Greenleaf 2016, 2–3; Leroy 2016, 3).

We will now illustrate the above arguments in practice through a focus on how the TiSA would balance three prominent values, i.e., free trade, privacy, and security.

Free data traffic is the general rule of the *Annex on Electronic Commerce* of the TiSA, as a corollary of the wider principles of National Treatment and Most-Favored-Nation Treatment, provided in the Core Text (Articles I-4 and the un-numbered Article between I-2 and I-3). Article 8 of the *Annex on Electronic Commerce* forbids location requirements for computing facilities and Article 2 mentions the right of 'transferring, accessing, processing or storing information, including personal information, within or outside the Party's territory, where such activity is carried out in connection with the conduct of the service supplier's business'. Here 'connection' is a broad term able to account for all of the different hypotheses where economic activities require the transfer of data: data can be an accessory, the object or even a remuneration for the service.

The same Article 2 of the *Annex on Electronic Commerce* unreins the circulation of personal data, too. In this regard, the data subject's interest is somehow recognized but not entrenched with specific safeguards. More precisely, privacy is mostly protected as a consumerist entitlement, instrumental to 'consumer's confidence in electronic commerce' (Article 4, *Annex on Electronic Commerce*). Consequently, a real, albeit weak, consent requirement is only set for unsolicited commercial electronic messages (Article 5, ibid.). In the remaining fields, there is only a broad-termed obligation: the Parties must adopt a 'domestic legal framework' of protection (Article 4, ibid.).

In practice, the Parties can have privacy legislation. However, its approval or enforcement must not alter the substance of the free economic initiative. As anticipated, any conflict that may arise between the two values is settled through a rule-exception mechanism, where privacy is an exceptional norm, and can derogate to free trade only if very strict requirements are met (Greenleaf 2016, 4–5; Irion, Yakovleva and Bartl 2016, 29–38). Pursuant to Article I-9(c) of the Core Text, data protection measures shall not be a 'means of arbitrary or unjustifiable discrimination between countries where like conditions prevail, or a disguised restriction on trade in services' (Article I-9, chapeau, Core Text), shall be 'necessary' and 'not inconsistent with the provisions of this Agreement'.

For some parties' systems, it is difficult to predict whether the implementation of data protection law would pass such legality test.

An example is the European Union. Therein, the *General Data Protection Regulation* (or GDPR, [EU] 2016/679)—which on May 25, 2018, is substituting the *Data Protection Directive* (95/46/CE)—forbids data to be transferred to third Countries that do not ensure an 'adequate' level of protection (Articles 44–47). Moreover, that 'adequacy' has been rigidly interpreted by the EU Court of Justice. Indeed, this Judge reads the requirement as imposing an 'essentially equivalent' level of protection[11] (De Minico 2016, 218–28). Now, the possibility to enact such a policy would actually be doubtful under the TiSA regime, because every implementing act would have to fit the indefinite language of the Article I-9 derogatory clause. A confirmation comes from the jurisprudence under Article XX of GATT and Article XIV of GATS, whose wording is similar to the Article I-9 of TiSA. In some instances, the exercise of the State authority to legislate in favor of fundamental values has been ruled unlawful[12].

Moreover, we know little about the enforcement organisms and proceedings of TiSA, which are barely sketched in the latest draft (*EU Proposal, TiSA—Dispute Settlement Chapter*). This is a second point of friction with some Parties' constitutional systems: in the same EU example, Articles 51–58 of the GDPR provide a due process clause and certain conditions of independence and neutrality of the deciding organism. As explained in the next paragraph, this difference is not anecdotal since the author and the procedures of the decisions also determine its contents.

Given all these circumstances, it is probably true—as EU Institutions claim[13]—that TiSA is not going to change the parties' data protection laws. But it is also true that these privacy rules risk being voided in their enforcement possibilities because of the broad and numerous limitations provided (Kelsey 2016, 2; Weber 2012, 32–34).

Security is also carved as an exception to free trade. However, it is a stronger exception compared to privacy, given that in the relative norm (Article I-9(a), Core Text) there is no mention of the requirement 'not inconsistent with the provisions of this Agreement', which, conversely, is applied to privacy. In addition, in the cases falling under the specific 'security exceptions' listed in Article I-10, the terminology of Article I-9, chapeau, is not replicated. Therefore, the only remaining limit is that derogation ought to be 'necessary' (Kelsey and Kilic 2014, 16). In this way, not only does the TiSA prioritize market concerns over people's safety, but also sets a discretional hierarchy between two fundamental and often competing rights.

In conclusion, TiSA, if left unchanged in its final version, would assume free trade as an all-encompassing value, subject to few bounds deriving from basic liberties. Therefore, the practical effects of the Agreement would be entirely entrusted to the bodies and procedures of dispute settlement established within the TiSA. But, for the reasons that are to be explained in the next paragraph, this aspect is still fuzzy and problematic in the Agreement.

INSTITUTIONAL TORSIONS OF FREE TRADE AGREEMENTS

'The role of market exchange as a coordination mechanism is a defining feature of capitalism. But its scope can and does vary across different accumulation regimes and modes of regulation. It has the greatest weight in liberalism and, more recently and even more extensively, in neoliberalism. This is not limited to the coordination of economic relations narrowly understood but extends to political relations and the nature of civil society' (Jessop and Sum 2006, 258).

A path that must not be read as a symptom of the States' sunset, but, more appropriately, as a proof of their active role, which makes them co-creators of the economic institutions and practices as well as '"power connectors" in networks of states plus non territorial forms of political organization—networks that reflect and refract the balance of forces in their respective political space' (Jessop 2016, 192).

The institutional aspect of the TiSA can only be understood if we cast a background analysis of the transnational multilevel systems of regulation. These have been described as a form of 'new constitutionalism' because, through the growing use of legal system, they are designed to lock in neoliberal policies and thus insulate them from democratic influence (Gill 1995;

Gill and Law 1988). A vision which is certainly very distant from the idea of modern constitutionalism (McIlwan 1947; Fioravanti 2009) but is fed by the fact that these highly institutionalized systems of commercial exchanges 'should receive treatment more analogous to the way in which national institutions (including courts) treat a "constitution," rather than the traditional way of treating treaties' (Jackson 2006, 51). With regard to a different Trade Agreement (the *Anti-Counterfeiting Trade Agreement*—ACTA) it has been observed that it is clear that the development of new regulations, often under the shadow of governments, tends to escape the control of democratic and legal institutions. Most often it is a question of sophisticated implementation technologies discussed by closed groups of high-level bureaucrats, engineers and public and private decision makers, whose compliance with democratic and legal norms is difficult to assess (Brousseau and Marzouki 2012, 388).

In the latest generation of FTAs, of which the TiSA could become part, this form of circumvention seems to be making a quantitative leap. The governance bodies that are going to be instituted are intended to assume solid and expanding para-normative and para-jurisprudential functions, that are mainly borrowed from the Investor-state dispute settlement system (Micciarelli 2017, 257–61; Ferrarese 2014; 2017). A model tailored on a new diplomatic generation, made up of negotiators and advisors chosen among 'experts'— often co-opted by private businesses and multinational enterprises—that incarnate de facto the academy of a new diplomatic establishment.

Now, it is undoubtable that the problem of procedures and accountability is central in every possible form of decision-making; according to a minimum definition of democracy 'each rule does not impose what to decide, but only who shall decide and how' (Bobbio 1999). Also in this case, the 'what' largely depends on how the rules set the 'who' and the 'how' of the deliberations. And it is exactly here that a deep divide between 'old' and 'new' institutions is generated. And this appears very distant from the system of safeguards and public involvement typical of the democratic-representative circuit.

We can only list, briefly, some examples. First and foremost, the problem of secrecy in the procedures. The TiSA negotiations are not merely informal, but indeed secret[14]. Democracy is, by definition, the government of the public good. A promise that, not later than twenty years ago, was considered one of the most unkept promises of democracy (Bobbio 1999, 363). The difference, in this case, is that secrecy is not due to the traditional 'Reason of State', which calls to protect national security. Rather, it serves the 'reason of poker': it would be unwise to reveal one's own cards to one's opponent in the middle of a game.

Secondly, the scarce official documentation which has been disclosed seems to be shaped by an approach that we could define as 'institutional populism'. We can consider as an example the EU position papers and fact sheets that claim, for instance, in colored *dépliants* that 'Nothing in TiSA would stop a country from applying its confidentiality or data protection laws'[15]. A vow that, as we have seen in the precedent paragraph, is doomed to remain unaccomplished. This causes a distortion also in terms of political communication: governments cannot provide reassurance, if not in a populist way, since these agreements will live their natural life, will be interpreted by other organisms, courts, and committees of experts, and will be applied in trials, which is to say arbitration proceedings that will count as precedents. In short, they will develop a whole enforceable juridical system characterized by a private matrix.

From this perspective, what is coming to light is a systemic model of norm production, which is 'autopoietic' and 'reflexive' (Teubner 1993, 2012). All this does not amount to a mere lobbying, or an extension of the US model (where lobbies legitimately design broad public policy choices (a perspective that would still be unprecedented, because here the transnational dimension would make it even more difficult to control the actors). It rather constitutes a step ahead, namely the redefinition, also formal, of these powers: not a mere *influence* of stakeholders and pressure groups, but a *restructuring* of institutional architectures and logics of governing, whose political shape is tailored on these private subjects' organizational structure. In this way, free trade governance not only favours those imbalances, but also reshapes the rulemaking, by privileging the direct intervention of corporations and stakeholders over popular control[16].

At first glance, the free trade governance appears to be an ideal model, as it succeeds in providing a platform of permanent negotiation and consultation to public and private interests. But the pledge to build a horizontal and networked model seems to conceal a hierarchy immanent in the very social ground where such participation is rooted: i.e., economic relations (Tucci 2012). A ground where intervening with a political and conflictual action is much more difficult, because it is difficult to create institutional bodies for an effective participation of *all* the interested persons: weak people, unorganized consumers, spontaneous activists, etc. As Marx wrote, 'Between equal rights, force decides', and this mirrors the ambivalence of the horizontality of the network. Indeed, we have observed earlier the corporate appropriation of the netizens' collective 'digital labor', which shows that hierarchies arise, paradoxically, in inter-private relations taking place on a horizontal infrastructure (De Tullio 2016, 641–51).

CONCLUSIONS: GOVERNAMENTALIZATION OF INSTITUTIONS TOWARD FREE TRADE GOVERNANCE

Today, in an interdependent world, the market structures that regulate supply and demand seem to offer one of the few truly global institutional models that is also one of the most unfair and unbalanced ones.

It is true that institutions keep evolving continuously, despite the fact that their role is to stabilize the political system. Free trade governance is a paradigm of neoliberal institutional design because it combines economic rationality with a political rationality, thus breaking every fine line between the two political and economic spheres. We could state that free trade governance unveils *a tendency, a line of force* of the process that Michel Foucault has called governamentalization. Governmentality is a profound modification of the reason for government, the dynamics of which are to be located in the regulation of populations and the forms of power that target them. Foucault shows how the lexicon and objectives intended to guide government action in a process that has been developed since the fifteenth and sixteenth centuries, are no longer directed towards the pursuit of the 'common good', but rather towards that of the 'convenient end' for each of the items to be governed and clearly borrowed from an economic and utilitarian logic. Techniques of government over the population were developed, inseparable from the truth discourse of specific forms of knowledge, like biology, psychiatry, and economy (Foucault, 2004; 2000).

Governmentality studies have turned our attention to the issue of governing, showing how power techniques directly address the transformation of state and other institutions as well as the transformation of the individuals in a connected process of subjection and subjectification (Miller and Rose 1990; Burchell, Gordon, and Miller 1991; Lemke 2002). Governmentality helps us to read the core of power trajectories crossed by the encouragement of self-government individuals' choices and preferences. Thus, neoliberalism has been successful in remodeling the state's institutions and authorities according to its rationality (Dardot and Laval, 2015), and in creating a new institutional order that has a private normativity which comes from the world of private powers (Sassen, 2007, 40). Moreover, a real revolution is also a socio-economical one, fed by the osmosis between the plane of institutions and that of the production and exchange of goods and services, precisely because the latter involves billions of people, capitals, and physical and virtual technologies in 'a mixture of economy and technique that finds in itself its own operating logic and in the living bodies the point where power is enforced' (Bazzicalupo 2006, 108). The neoliberalization of society coincides not only with the institutional revolution, but also with the extension of the evaluation of social life according to economic-monetary criteria. In this sense, 'the real novelty of anthropological significance introduced by neoliberalism is precisely the tendency to merge the market and life into the same paradigm' (De Carolis 2017, 23).

The world of goods and services—that in the capitalistic order also means the world of labor, rights and duties, psychopathologies, desires and self-fulfillment—clarifies the connection between the material, economic, and ideological planes which overcomes the Marxist dynamic of structure and superstructure, showing how it 'should be seen not as three levels, but as analytically distinct, coterminous moments both of concrete social practices and of concrete analysis' (Garnham 2006, 205). And this clarifies the necessity of focusing 'on institutions and governance as they are both imagined and practiced. How, for instance, are social imaginaries invoked by different models of governance? What moral order is constituted concerning the rights and obligations we have to each other?' (Mansell 2016, 22).

Questions that are as pertinent as they are complex, and on which we believe it is possible to find—in the dialogue between Internet governance scholars—a common field for reflection. The Internet and the rights and interests around data flow today seem primarily functional to support preferences and desires according to the rationality of the capitalist market, in its neoliberal version. But the Internet is much more: it is a formidable platform of interdependence that can also connect in different ways and forms and will be able to do so only if its governance is able to follow different multilateral institutional paradigms.

NOTES

1. The work is the outcome of a shared reflection. First, third and fourth sections are written by Maria Francesca De Tullio; Second, fifth and sixth sections are written by Giuseppe Micciarelli.

2. http://ec.europa.eu/trade/policy/in-focus/tisa/.

3. European Parliament Resolution 2015/2233(INI), rec. B.

4. Instead, interdependence can be interpreted in other ways, as for example is shown by studies on the commons (Ostrom 1990; Micciarelli 2022).

5. This is also partly applicable to CETA, which entered into force on 21 September 2017, but only on an interim legal basis due to the hermeneutic conflict over its ratification, which should be read precisely as a conflict over the torsion of democratic-representative institutions produced by free trade governance.

6. A treaty whose ratification was revoked with one of the first executive orders of the newly elected President Trump. That is the certification of a crisis scenario for neoliberalism, which shades uncertainty and spreads debate not only over the *iter* of approval of the TTIP, but also over the future of these Treaties.

7. In that sense, see *Proposal for a Regulation of the European Parliament and of the Council on European data governance (Data Governance Act)*, COM(2020) 767 final, 2020/0340(COD), Brussels, 25.11.2020 (EU Commission 2020), as a part of a the broader EU data strategy: https://ec.europa.eu/info/strategy/priorities-2019-2024/europe-fit-digital-age/european-data-strategy.

8. The risks of this situation are addressed in a recent proposal which provides more rights for newcomers, even if it does not radically question these market concentrations: *Proposal for a Regulation of the European Parliament and of the Council on contestable and fair markets in the digital sector (Digital Markets Act)*, COM (2020) 842 final, 2020/0374(COD), Brussels, 15.12.2020.

9. *Ex multis* European Court of Human Rights, *Klass v. Germany*, Application no. 5029/71, para. 42; paras. 66–8, 81.

10. *Ex multis* European Court of Human Rights, *Halford v. the United Kingdom*, Application no. 20605/92, paras. 66–8, 81; European Court of Justice, *Digital Rights Ireland Ltd c. Ireland*, joined cases C-293/12 and C-594/12, para. 38.

11. European Court of Justice, *Maximillian Schrems v. Data Protection Commissioner*, C-362/14, para. 73.

12. WTO Panel, *United States—Measures Affecting the Cross-Border Supply of Gambling and Betting Services*, WT/DS285/R, 2004; *WTO Appellate Body, United States—Standards for Reformulated and Conventional Gasoline*, WT/DS2/AB/R, 1996.

13. http://ec.europa.eu/trade/policy/in-focus/tisa/questions-and-answers/.

14. A secrecy which is not limited to the phase of negotiations. In one of the texts published by WikiLeaks, i.e. the first page of the Annex X—Financial Services (April 14, 2014), a disclaimer is very clear about the terms of declassification: 'Five years from entry into force of the TISA agreement or, if no agreement enters into force, five years from the close of the negotiations'.

15. http://ec.europa.eu/trade/policy/in-focus/tisa/questions-and-answers/.

16. This impression is not attenuated by the consideration that these procedures are formally open to 'citizens and stakeholders'. Indeed, in substance, the decisional *iter* is modelled on the latters' structure, and the only subjects of civil society that are equipped for such a vast political engagement on transnational scale are the biggest NGOs, trade unions and recognized associations. This is, indeed, a forceful limit to the spontaneous participation of citizens, that only leaves the floor to a format similar to opinion polls conducted with the system 'your voice in Europe', http://ec.europa.eu/yourvoice/consultations/index_en.htm.

Chapter Nine

National Sovereignty, Global Policy, and the Liberalization of Telecommunications Markets

Claire Peters

Lawful interception is the surveillance of a designated target or targets over communications networks as permitted within the scope of the law. The phrase is commonly used in the context of policing and law enforcement agency (LEA) activity, which may or may not concern itself with national security issues in addition to police work. As pen registers were rendered obsolete and references to 'wiretapping' became inaccurate without wires to tap, 'lawful interception' became an increasingly common term in the digital age for what were once diverse warranted surveillance procedures governed by different sets of laws. With the advent of Internet and satellite services, the meaning expanded to include the collection of even broader kinds of information. The phrase now encompasses any legal method for the targeted interception of data and metadata.

On 17 January 1995, the Council of the EU passed a council resolution (1995) signifying a nonbinding agreement among EU member states to adopt designated 'International User Requirements' (IUR) for telecommunications companies. The IUR stated the responsibility for telecommunications service providers to supply 'a real-time, fulltime monitoring capability for the interception of telecommunications' (Council of the EU 1995: 2) when an LEA presents the appropriate warrant. Although many member states already had similar requirements for private operators, the IUR provided a common framework with which to standardize LEA-service provider interaction throughout the EU. Under these policies, all service providers operating in a state are required to set up a handover interface for the transmission of communications to a law enforcement monitoring facility (LEMF); be capable of collecting and storing the content and metadata of a target's call or connection; be able to either decrypt or provide the means to decrypt any communications they have encrypted; and transmit, on demand, the requested content and associated metadata for a specified target's communications to the LEMF.

The 1995 council resolution was paralleled by the adoption of similar laws in the US and Canada that same year, and in Australia in 1997. In the following decade, additional countries would use the IUR as a point of reference for their own laws. Some, including New Zealand and South Africa, adopted bills that amended their telecommunications regulations to designate similar expectations and encourage design-stage planning. Others, such as Kazakhstan and Egypt, instituted IUR-style requirements for private companies in the same bills that established regulatory authorities for their own newly liberalized markets. The IUR became the basis for technical standards developed by the European Telecommunications Standards Institute (ETSI) that provided interfaces for the interception of Internet services and mobile telephony, among others (Hosein 2013). The ETSI standards were later endorsed by the International Telecommunications Union (ITU) as a global benchmark for interception standards (Dempsey 1997). The ubiquity of the IUR and the availability of global standards that fit them has caused telecommunications companies to design their equipment and services to their specifications as a matter of course.

A growing body of Internet governance scholarship examines government reliance on private companies for the pursuit of intelligence and security-related goals. A number of papers (e.g., Musiani 2013; Arpagian 2016) raise concerns about the threat that current public-private cooperation on surveillance poses to human rights. Elkin-Koren and Haber (2016) voice these concerns from a legal standpoint in their examination of the 'legal twilight zone' in the United States that encourages informal private-sector cooperation with government surveillance—a phenomenon the authors label 'governance by proxy'. This chapter aims to contribute to this important conversation through a socio-historical analysis of the IUR, examining the conditions and considerations that shaped their development and substance. As widely imitated policies that have influenced technical standards and private-sector norms, the IUR have enabled the international expansion of governance by proxy as it relates to surveillance. Analysis of their origins and effects reveals the deep connection between their genesis and the privatization of telecommunications, the influence that heavily bureaucratized LEAs had on their development, and the ways that the privatized telecommunications sector remains affected by their passage.

The EU and its member states serve as the objects of analysis because they played central roles in the international institutionalization of lawful interception policies and their actions were motivated by diverse incentives—collective interest in regional integration informed the EU's actions, as did the individual security interests of its member states. The varied positions and concerns of European countries enable an examination of surveillance policy that incorporates issues of governance diffusion and interdependence—a perspective which is more difficult to distill with a single hegemonic power like the United States as the focus.

Section 1, 'Lawful Interception in Policy and Practice', relates lawful interception policy structure to the mutability of the security state in a globalizing environment. It establishes a basis for understanding the IUR as a means by which member states negotiated privatized, transnational telecommunications networks' disruption of security-related activities. The liberalization of telecommunications markets replaced reliable state-run telecommunications agencies with widely varied actors and chaotic technological development, compromising security agencies' interception strategies. This threat prompted the introduction of policies that standardized and routinized private-sector provision of interception assistance across a variety of telecommunications networks. The international institutionalization of the IUR can therefore be analyzed as a form of security state adjustment to the diffusion of power to the private sector.

Section 2, 'Lawful Interception and Police Bureaucratization', elaborates on the basic premise of security state adjustment by focusing on the design and passage of the IUR, which was characterized by secrecy and deceit. This secrecy can be understood as a byproduct of police bureaucratization—bureaucratic autonomy and the nature of their organizational mandate encouraged LEAs to exclude non-state actors, under the presumption that ensuring state power will preserve security. However, the principle that it is the right of security organizations to be technically capable of accessing all communications has not only heightened potential for the abuse of state power but also sacrificed absolute gains in infrastructural security for relative gains in national security.

Section 3, 'Observing Systemic Effects', examines the effect of IUR-based governance by proxy in the private sector. The private sector is not in a good position to balance LEA interests—small companies especially are at a disadvantage if they wish to challenge a government ruling, and profit incentive encourages a compounding of government and private-sector interest in the capture and analysis of personal data. Companies have become more conscious of the ethical issues that arise in their interactions with governments, but corporate social responsibility alone is not likely to serve as an adequate solution.

LAWFUL INTERCEPTION IN POLICY AND PRACTICE

The security state—that is, a state which holds considerable power over its citizens but operates under a mandate of ensuring their safety—is not exclusively concerned with its own existential security. Because the provision of security for a society includes protection from hardships experienced at the individual level, social and economic security have become vital to the security state's understanding of the 'national interest'. Much social and economic security has proven easier to pursue with the incorporation of privatized and international approaches. Security states are therefore incentivized to increase economic interdependence and delegate some functions to non-state actors. The diffusion of state power, in turn, shifts the contexts within which their security strategies operate, requiring states to either adapt to their circumstances or alter them (Mabee 2009). The evolution of lawful intercepts in policy and practice exemplifies this trend. The liberalization of telecommunications markets ended the implicit coordination of telecommunications services with LEA needs, thus compromising the ability to ensure state interception capabilities through national legislation. In response, EU member states collectively reassigned the responsibility for ensuring intercept capabilities to telecommunications companies.

During the 1980s, many countries enacted telecommunications privatization programs in efforts to incite economic growth and technological development. European countries were interested in adopting this strategy—market competition was considered an antidote to the slow, lumbering gait of state-led innovation, and some national communications providers expressed overt interest in the opportunity to expand into international markets (Rodine-Hardy 2013). As a regional entity, the EU had its own motivations—exchange among European telecommunications sectors would contribute to the goal of a common regional market (Schneider 2002). A series of directives were incorporated into member states' national laws and introduced competition into different market segments, starting with equipment in 1988 and ending with basic voice services in 1998 (Pauwels and Delaere 2007).

Official stances on lawful interception over much of the twentieth century were shaped by three underlying conditions, all of which the privatization and internationalization of telecommunications services would disrupt. First, telecommunications service provision primarily occurred at the national level. Before 1980, telecommunications was widely considered a natural monopoly due to the economies of scale and density involved in the maintenance of large communications networks. For this reason, European countries typically had a national Post, Telephone and Telegraph agency (PTT) which handled all major communications services of the day. PTTs were responsible for all aspects of a telecommunications network, from service operation to equipment manufacturing (ibid.). The coordination of services by large, stable bureaucracies ensured order and consistency in the provision of intercepts.

Second, the nationalized character of telecommunications naturally resulted in wide variation among national interception policies. In France, for instance, police interpreted general statutes on privacy and investigation powers as they saw fit until 1991, when the European Court of Human Rights compelled them to make a more coherent policy framework (Galli 2016). In contrast, West Germany introduced comprehensive legislation on wiretaps and other forms of intercept in 1968. West German attitudes toward interception were shaped by a strong belief in the right to privacy on one hand and heightened national security concerns on the other, providing early incentives for clear legislation (Carr 1981). Police in the Netherlands, who enjoyed a high level of trust from the general public, could afford to be more informal in requests for and utilization of intercepts (Klerks 1995).

Third, national lawful interception policies built on the assumption that LEAs were always capable of eavesdropping, and that the purpose of legislation was to designate when agencies were allowed to eavesdrop. Communications have traditionally been insecure by nature. Lawful or unlawful interception was therefore possible through the exploitation of any number of design aspects. The method preferred by LEAs was to commission the network operator to bridge the target's line to a friendly circuit with a loop extender, but do-it-yourself options were abundant (Landau 2011). Bugs could be inserted into a phone receiver, wires could be tapped on-site——legislation on lawful interception was passed to outlaw unsanctioned use of interception and constrain police investigations to acceptable limits, not to ensure government access to communications. The ease of interception, combined with the advantages that come with access to privileged information, could even cause strong interception restrictions to backfire. The complete outlawing of wiretaps in the US between 1934 and 1968 simply caused the FBI to conduct intercepts in secret, counter-intuitively enabling the expansion of such operations (Theoharis 1992).

The conditions affecting LEAs' use of intercepts were drastically changed by the opening of telecommunications markets. How they were challenged can be observed in the evolution of European mobile phone networks. Liberalization policies' primary objectives—rapid, free-form technological development and the expansion of markets—were precisely the source of LEAs' technical difficulties. Many European countries, foreseeing potential complications, instituted legislation to ensure that private service providers would continue to comply with intercept requests. The United Kingdom's 1984 Telecommunications Act, for instance, included a section mandating service provider compliance with state orders; the Netherlands' 1994 Mobile Telecommunications Act delineated contracted service providers' obligation to ensure interceptability (Koops and Bekkers 2007); Germany made intercept capabilities a provision of its network license requirements (Berke 1992). However, national mandates for private-sector assistance with intercepts proved inadequate preparation for an integrated regional telecommunications market. Even if a contracted national network vendor complied with national interception standards, a resident could subscribe to a foreign European mobile phone service which obeyed different laws or did not have open lines of communication with that country's police. Adequate policies concerning telecommunications service providers' compliance with foreign interception orders did not exist (ibid.).

The fast-paced technological innovation promised by liberalization also proved problematic. In Germany, the software needed for interception had not yet been produced for the transmitter stations of the country's two mobile phone networks. Although this technically violated Germany's network licensing provision, the networks were already in use; German police were left with no feasible recourse (ibid.). Similarly, the Global System for Mobile Communication (GSM) standard, which was employed by all European digital mobile phone vendors, incorporated encryption for both logistical and security reasons. The ETSI, which had developed the standard, had neglected to build in a corresponding mechanism for interception (Wong 1995). Because creating interception methods for a standard that already exists is much more difficult than incorporating one at the design stage, retroactively designing a means of regulated GSM interception would take until the end of the decade (Piper and Walker 1998). The rate at which new technologies were being introduced, combined with the time it could take to design accompanying interception methods, threatened to routinely compromise intercept capabilities at every point of access. A given technology might already be rendered obsolete by the time a mode of ingress was established.

The development of new technologies and the convergence of old ones required a fast response—if privacy became the norm in new technologies, political resistance would be greater and technical enactment more difficult if governments later decided they needed access. Securing and expanding the surveillance capabilities that governments already possessed was therefore perceived to be the best option. Tailoring policy to specific technologies required more specialist knowledge and faster responses than governments could muster, so instead of designing policy in accordance with specific communications technologies they established regulations that applied to private communications services as a whole.

The IUR mediated liberalized markets' disruption of state security strategies through the deputization of service providers, international policy coordination, and blanket provisions for all telecommunications. By fully delegating to telecommunications companies the responsibility to ensure interception provision at the design stage, the IUR established a hard-line stance toward private companies—design intercept capabilities into the network, or do not go to market. International cooperation around the IUR resolved the issues that stemmed from disparate national lawful interception policies. Once member states' expectations of service providers were expressed in accordance with a common framework, companies would need to tailor their approaches only minimally across different countries. The adoption of the same policies by third states and the ITU's endorsement of ETSI technical standards also addressed concerns from private industry about intercept provision policies posing national or regional disadvantages in a global market (Hosein 2013). Finally, blanket coverage of all telecommunications guarded against contingency. By applying to a category of private actor instead of a particular technology, the council resolution provided a soft basis for the coordination of lawful interception policy in current modes of telecommunications as well as future developments.

The way lawful interception policy was reconfigured for the landscape of liberalized telecommunications markets is characteristic of a security state adaptation strategy which Mabee describes as 'greater internationalization for national autonomy' (2009: 14). Globalization impacts security provision 'in terms of shifting the organization of state power beyond the national level,' but the shifts are caused by 'a tension between state power and the transnationalization of power, and as such, a constant strand throughout has been to emphasize the continuing power of the state' (ibid.:141). European member states opened their telecommunications markets to increase technological innovation, economic growth, and regional identity, but they faced unexpected challenges to sovereign powers. By instituting policies that routinize service provider assistance in lawful interception, states reconfigured economic developments in ways that minimized the need to alter their security strategies.

LAWFUL INTERCEPTION AND POLICE BUREAUCRATIZATION

Although a security state acts in the perceived interest of its citizens, it inevitably pursues objectives related to its own existential security in ways that categorically favor state interests over those of non-state actors, including its own citizens. Looking only at the logic guiding state decisions as they have been outlined, the IUR might be perceived as compromise between state and non-state actors. After all, with telecommunications systems completely out of government hands, enlisting the private sector became the best—possibly the only—means available to ensure dependable access to intercepts. However, the IUR development process tells a different story. The IUR were designed in secret by an international collaboration of LEAs that blocked private, civic, and even non-security-minded state actors wherever possible in order to ensure minimal resistance to policies that effectively deputized private industry. To understand how the policies of democratic states can be shaped by what could be conceivably labeled antidemocratic processes requires the recognition of another means of security state transformation—the bureaucratization of state agencies, especially of police forces.

The modern security state is highly bureaucratized—through the division of labor into agencies that are granted a degree of functional autonomy, a state can effectively pursue numerous distinct but overlapping goals (Deflem 2002). Facing problems like transnational crime which are best addressed through collective action, LEAs are incentivized to pursue objectives through international police cooperation when they cannot be achieved through national processes (Deflem 2002; Gerspacher 2005). The usefulness of international cooperation and the importance of police work to a state's existential security have caused present-day LEAs to reach 'a level of bureaucratic autonomy that is unprecedented in scale' (Deflem 2006: 244). Calling back the 'tension' referenced by Mabee, LEAs coordinate among themselves with significant autonomy from their respective states to do what they deem necessary to preserve sovereign self-determination.

The origins of the IUR lie in the FBI's rejected 1992 draft bill for 'Law Enforcement Requirements for the Surveillance of Electronic Communications', which required telecommunications providers to possess facilities for the provision of 'a real-time, fulltime monitoring capability for intercepts' (FBI 1994: 5). In 1993, the FBI quietly invited representatives from LEAs in Canada, Australia, Hong Kong, the UK, Denmark, France, Germany, the Netherlands, Spain, and Sweden to meet in Quantico, Virginia, to discuss new lawful interception policies that would address the changing telecommunications landscape. At the first of these meetings, dubbed the International Law Enforcement Telecommunications Seminar (ILETS), the FBI proposed their law enforcement requirements as a policy framework. Participants agreed on a slightly modified version of the FBI's original proposal in Bonn the following year (Campbell 1999). This framework became the basis for the EU's 1995 European Council Resolution, the United States' 1995 Communications Assistance for Law Enforcement Act, Canada's 1995 Solicitor General's Enforcement Standards for Lawful Interception, and Australia's 1997 Telecommunications Legislation Bill.

Activities connected to the IUR were consistently characterized by evasiveness and opacity. Approved through telex exchanges and not published in the Official Journal of the European Union until November 1996, the January 1995 council resolution bypassed the scrutiny of the European Parliament, national legislatures and the wider public (Medosch 2001; Koops and Bekkers 2007). Even as the resolution was incorporated into member states' national legislations, ILETS and the FBI's role in its design appeared to be deliberately omitted (EPIC 2002: 34; Medosch 2001). In November 1995, the fifteen member states and Norway signed a Memorandum of Understanding (MOU) inviting third States to cooperate on IUR-based policy. Canada, the US, Hong Kong, Australia, and New Zealand confirmed their intentions to incorporate the requirements into national law and participate in related cooperation (Bunyan et al. 2000). Through the MOU, ILETS was rebranded as an informal working group through which member states and third States could collaborate on interception-related issues (ibid.).

Following their establishment as a working group, ILETS indicated their intention to expand surveillance operations with a new draft council resolution on the Lawful Interception of Telecommunications in relation to New Technologies. The draft, labeled 'ENFOPOL 98', included extensive specifications relating to different forms of communications. For Internet interception standards it included requirements for user password and email logging, key recovery, and so on. Designed solely within LEA-headed working groups, ENFOPOL 98 became the subject of public and Parliamentary debate after a draft was leaked to the press (ibid.). The matter was finally brought before the European Parliament when evidence emerged linking the specific requirements of ENFOPOL 98 to UK-US plans to expand 'ECHELON', a military surveillance program, and still more evidence indicating that ECHELON had been used for industrial espionage on EU member states (Schmid 2001). While shock and anger over the revelations resulted in the shelving of the ENFOPOL 98 specifications, similar requirements were enacted over the following years—especially after 11 September 2001.

Characteristics of the LEA mandate that inform the nature of their autonomous activities explain the route that international collaboration on the IUR took. First, as a matter of course, bureaucratic autonomy causes some divergence in perspectives and objectives between LEAs and their national governments. Aldrich (2009) and Deflem (2002) show that the behaviors of sovereign agencies in security-related cooperations indicate a perspective that adheres to a Westphalian security paradigm, even as their national governments demonstrate earnest interest in interdependence. Even in international security cooperations that yield absolute gains, agencies' actions betray a sense that the long game remains zero-sum. This zero-sum perspective ensures LEAs put their own states first even as autonomous entities, but this also encourages some amount of mistrust of any organization outside their state's hierarchy. For instance, international relations and security theorists have noted a hesitance inherent in security-related cooperations, even when international cooperation is clearly the best way forward (Deibert and Rohozinski 2010; Mabee 2009). In the case of interception, any entity that devalues state security—private industry and civil society is perceived as a threat. By rechanneling private power back into the state and limiting non-state actors' means of protest, the position of sovereign actors is reasserted and the threat removed.

Second, LEAs' mandate of ensuring a society's safety through the prevention and prosecution of criminal acts makes them more likely to conflate true security with their own operational success and overemphasize their need to gain a competitive advantage in their work. Dragu's (2011) game theory analysis of post-9/11 LEA attitudes toward privacy indicates that LEAs are incentivized to reduce privacy protections even when it stands to make society more unsafe, because it always benefits their expected outcome. It is conceivable that LEAs used secrecy out of a sense of urgency—that their weakening powers were genuinely perceived as an emergency. As in a state of national security, therefore, they took extraordinary measures, choosing to avoid wasting time in a protracted public debate and instead expedite the securing of intercept capability. Unfortunately, this secrecy also enabled the process to be hijacked by US political interests.

Finally, LEAs' bureaucratic autonomy and conflation of power and security pit network security and national security interests against one another. The IUR represent a significant change in state approaches to lawful interception policy—they ensure, rather than restrict, intercept capabilities. Surveillance practice has therefore moved from exploiting system features to requiring exploitability as a system feature. This puts the security of virtual communications infrastructure and the interests of domestic security actors at odds with one another to an unprecedented degree (Deibert and Rohozinski 2010). Expert consensus warns against the intentional incorporation of security vulnerabilities into global networks (Abelson et al. 2015; Diffie and Landau 2009). The trouble is that it is easier to justify pursuing the quantifiable gains of interception ability over the unquantifiable and unattributable costs that eventually result from the compromise of infrastructural integrity. The true cost of damage to infrastructural integrity only reveals itself once something goes disastrously wrong—for example, the 'Athens Affair' of 2004, in which an unknown party hacked into the built-in intercept architecture of Vodafone networks and eavesdropped undetected on public officials and human rights activists for months.

It is difficult to miss how LEA-designed policy pits public order against privacy even though LEAs' ultimate goal—the security of their states—is not actually weakened by a right to privacy. The juxtaposition of order against privacy is not ineluctable, but a result of policymaking being dominated by LEAs' organizational perspective, and operating in accordance with it does not necessarily make society safer. The effects that policies advocating governance by proxy have on the private sector further proves the importance of looking beyond immediate benefits.

OBSERVING SYSTEMIC EFFECTS

Even if intercept capabilities were proven to be a boon to LEA or national security operations, those benefits must be considered alongside the policies' systemic effects. 'Governance by proxy' is prone to exploit the for-profit nature of the private sector, which makes it difficult for telecommunications companies to check government powers and tempting to profit from magnifying them. These policies are therefore rendered nearly invisible, as attention is turned to the tangible actions of private industry and away from the underlying forces that encourage them to act in particular ways. This reassignment of responsibility has caused some positive developments, namely an increase in private-sector pushback to surveillance overreach, but the origin of the problem in government policy means more formal solutions are likely needed as well.

The member states' objective in passing the 1995 resolution was to ensure intercept capacities similar to or greater than those of the old LEA-PTT relationship. Although the desired capacities have been restored, the public-private partnership through which it is maintained is fundamentally different from the former inter-agency relationship. Companies possess more information resources and less legal representation than nationalized service providers. While agencies are tasked with civic duties and their existence is not necessarily threatened by a lack of competitive advantage, private industry's primary concerns are naturally oriented toward profit, expansion, and survival—not helping or serving citizens. A company which cares about the latter to the detriment of the former is less likely to continue existing.

The heightened existential risk faced by private actors and this risk's connection to financial performance is largely at odds with the private sector's presumed ability to check state power. On average, one can expect detached compliance with potentially dubious orders; at worst, companies find ways to capitalize on their circumstances by cooperating with LEAs beyond the degree required by law and producing additional services for their use (e.g., AT&T's Project Hemisphere). The public-private partnership supporting current interception practices therefore extends state power while removing checks to its execution. The international standardization of lawful interception policies has made private-sector provision of intercept capability a requirement akin to a terms of service agreement—it might technically be a choice, but a company cannot proceed as planned without agreeing, and little room exists for discussion or negotiation. Companies must therefore be willing to either comply with all the demands a country will make on them or not engage at all—as Verizon CEO Lowell McAdam commented when asked to speak about his company's controversial cooperation with the NSA, '[w]e are the largest telecommunications provider to the United States government, and you have to do what your customer tells you' (Alicea 2013).

The link between policy and the world it has engineered sometimes appears easy to forget. A certain dance is often performed in public debates over public-private surveillance partnerships. In the first act, a policy is proposed that expands surveillance power through private industry mandates. Civil rights groups bring up concerns about the compromise of information security, degradations of civic liberties, and the temptations of public-private collusion; in response, advocates of the bill brush these concerns aside by describing ways the rule of law will ensure the policy is used as advertised. In the second act, after the bill has become a law and opponents' predictions have come to pass, the problem is identified as corporate misbehavior and the conversation is almost exclusively concerned with matters of corporate social responsibility; the policy discussion has faded from memory.

In 2007, Iran had begun to privatize its telecommunications industry and was looking to attract foreign investors (Chussudovsky 2008). It successfully negotiated a mobile network contract with Nokia-Siemens, which was looking to penetrate new markets (Schrempf 2011). As part of the contract, Nokia-Siemens sold ETSI-compliant LEMFs to Iran (Rhoads and Chao 2009). The following year, nationwide protests in Iran erupted over suspicions of elections tampering. Iranian protestors used the Internet for organizing, to the degree that the United States petitioned Twitter to delay scheduled maintenance in a show of support. Ahmadinejad's government responded with the implementation of surveillance and filtering mechanisms, as well as a violent crackdown on online and offline dissent.

When a *Wall Street Journal* report traced a connection between Iran's online surveillance abilities and the installation of Nokia Siemens LEMFs the year before, the company experienced widespread criticism. European Parliament went as far as to condemn Nokia-Siemens by name in a 2010 resolution commenting on Iran's human rights violations; a jailed Iranian journalist and his son filed a lawsuit against the company (Watson 2010). Barry French, Nokia-Siemens' then-Head of Marketing and Corporate Affairs, was given an opportunity to speak about his company's actions at European Parliament's 2

June 2010 hearing on New Information Technologies and Human Rights. In the course of his statement, he identified an 'inherent tension' in the situation:

> Consider the fact that the systems that we provided to Iran were designed to implement a right that the ITU has explicitly said is held by Member States, and are required by law in the vast majority of those Member States. Yet, when we help to meet those requirements, we are subject to considerable criticism, including in the European Parliament resolution on Iran. (French 2010)

French is correct to highlight a contradiction in terms, as Nokia-Siemens behaved precisely the way established policy requires companies to behave— as a compliant assistant to LEAs. There is indeed an 'inherent tension' in the designation of private-sector complicity in government surveillance. It assumes that companies will somehow be in a position to recognize, more clearly than the states they serve, when they are aiding in human rights abuses. It also assumes that companies have the power to stop those abuses before they happen. Nokia-Siemens was internationally castigated for its actions due to some sense that they should have known their equipment would be horribly misused. In reality, such things are often only seen clearly in retrospect. Before 2009, Iran privatizing its telecommunications services and pursuing better ICT for citizens would have been positively interpreted by many as an empowerment of civil society and a sign of growing interest in peaceful integration into the international community (Ebrahimian 2003). Iran's rampant abuse of power is not the result of Nokia-Siemens' decision to do business with them; it is the result of policies and standards designed with the belief that the ability to abuse such power is one that states categorically deserve to possess.

The international standardization of LEA-service provider roles is designed to encourage a lack of concern from private companies regarding how the information they provide is used. Lawful interception laws are intentionally designed to shield companies from any consequences for fulfilling their legal obligation (Elkin-Koren and Haber 2016). Furthermore, principled opposition to a state demand is difficult. Opposition through the legal system requires a combination of resources and incentives which the majority of private companies do not possess (ibid.). Addressing the issue outside of the legal process is also a tall order—whistleblowing is, to understate the case, typically followed by disruption to one's own affairs. Given the incentive to cooperate with federal orders and the penalties for uncooperative behavior, it seems strange to require companies to act as both compliant service providers and protectors of civic life.

Of course, states are not the only actors capable of changing their behaviors. Calls for greater private-sector accountability have prompted telecommunications and tech companies to actively engage in a redistribution of culpability. Organizations like the Telecommunications Industry Dialogue and the Global Network Initiative promote corporate social responsibility around issues of privacy and freedom of expression; informal pushback against government demands are increasingly common (Elkin-Koren and Haber 2016). Nokia-Siemens' own business with Iran ended soon after the scandal, and the company divested its LEMF supplier in 2010. The call for accountability and transparency is one that companies, unlike LEAs, are in a position to answer. The formation of informal governance processes around the principle of corporate social responsibility has helped, at least in some small way, to fill a void.

Although corporate social responsibility shows promise as a potential counterbalance to overreach, it has its limitations. Western governments often emphasize the autonomy of the private sector when telecommunications and technology companies expand their services to non-Western countries, but demand trust, compliance, and assistance from the same companies when pursuing their own interests (Broeders 2015: 74–75). The use of Nokia-Siemens' surveillance technology to enable a sovereign state's violation of its citizens' human rights rightfully roused discussions over corporate social responsibility, but should have also revived debate over the policies which normalized private-sector assistance of interception. Especially with a growing sense of political instability in Western countries, the question should not be about who supplied these capabilities to Iran, but whether they should be supplied to any government at all.

CONCLUSION

The Internet as a 'disruptive technology' receives a lot of attention in Internet governance discussions regarding the problems of public-private collusion and governance by proxy, but the abundant precedent for these issues in telecommunications surveillance requires the origins of these issues to be traced farther back. The IUR, which represent a landmark in the instatement of governance by proxy as an international norm, are clearly linked to the liberalization of telecommunications. The unpredictable change it introduced disturbed agencies tasked with safeguarding social order, and therefore prompted policy that incorporated private-sector responsibility for intercepts.

However, allowing LEAs' conflation of power and security to be written into policy that now determines the nature of public-private interaction demonstrates a seemingly willful ignorance of the past. The boost of LEA capabilities that intercept capacity provides does not necessarily translate to making a more secure society. The public order v. right to privacy arguments that arise whenever digital-age surveillance becomes a topic relies on a false dichotomy that only serves to benefit LEAs and security agencies. Further-more, legal safeguards demonstrably fail to ensure that lawful interception will be the only kind that occurs. Architectures of control begin piecemeal— the technological basis for a Foucauldian surveillance regime being built up under one's nose cannot be averted through the power of good intentions alone. As Zimmermann (1999) observes, 'while technology infrastructures can persist for generations, laws and policies can change overnight.'

Chapter Ten

In This Bright Future You Can't Forget Your Past

Debating the 'Right to be Forgotten' in Latin America

Jean-Marie Chenou[1]

The Snowden revelations starting in 2013 came as a reminder for other countries that the historic stewardship of the internet by the US was not completely benign. Reacting a few months after the revelations, the then European commissioner for Digital Affairs, Neelie Kroes, analyzed the situation with a mix of value-based and geopolitical arguments. The necessity to defend 'European values' was made clear: 'Our fundamental freedoms and human rights are not negotiable. They must be protected online.' At the same time, she also insisted on the 'redrawing [of] the global map of internet governance' (Traynor 2014). As the values and geopolitical competition between the two sides of the Atlantic became institutionalized through new regulations and court decisions, Latin American countries discussed issues of privacy and freedom of expression, confronting two different visions of internet governance and content regulation. Indeed, while the United States continued to defend a private sector-led internet governance aiming at innovation and market growth, the European Union started focusing more on public regulation and a certain responsibility of (mostly US) private companies operating as intermediaries on the internet in the implementation of preexisting rights. The debate in the region was illustrated by the organization of the NETmundial conference by Brazilian President Dilma Roussef in April 2014. The final document of the multistakeholder conference recalled the importance of human rights and shared values on the internet, especially freedom of expression and privacy; and proposed a roadmap for the evolution of the internet governance ecosystem, echoing the declaration by the EU Commissioner (NETmundial 2014). At the same time, political discussions at the national

177

level focused on the responsibilities of internet intermediaries in the protection of fundamental rights.

Against this background, this chapter analyzes how legal expertise from the US and Europe fed the debates in Latin America regarding the liability of intermediaries. It argues that the power struggle between a European and a US vision in a context of crisis of existing regulations and values required the use of diplomacy by the two powers in order to influence the debates in the largest Latin American countries. The chapter draws upon the literature on norm diffusion and complements it with socio-legal studies in order to study internet law from a policy diffusion perspective. The chapter analyzes the debates on the 'right to be forgotten' in Latin America through the lenses of norm diffusion and thus focuses on the different strategies adopted by EU institutions and by US actors in order to export their own conception of the liability of intermediaries, and of the role of the state in the regulation of digital markets in general.

The remainder of the chapter is organized as follows. First, the next section situates the debates on internet policies as part of a norm diffusion diplomacy. Second, I describe the growing divergence between Europe and the US on the legal liability of internet intermediaries as a legal and political debate that epitomizes fundamental differences on how to grasp the legal challenges associated with the rise of cyberspace. Third, I outline the history of the debates on the right to be forgotten in the largest Latin American countries. Fourth, I explain the shift from a European perspective towards a more US-inspired approach in the debates on the 'right to be forgotten' in Latin America. Finally, I draw some conclusions on the differences between the two approaches to norm diffusion evidenced in the debates on the 'right to be forgotten' in Latin America.

INTERNET LAW AND NORM DIFFUSION

Internet law is not only a necessary framework for the development of online activities. According to Luhmann (2008, p. 146), legal norms are 'a structure of symbolically generalized expectations'. Law is essential in the anticipation of an uncertain future. As such, it is also a field of struggle among diverging worldviews and values that promote certain directions for the future. The field of internet law has historically been dominated by a US vision. However, other global powers such as the European Union are competing in order to diffuse their norms, values, and policies worldwide.

Norm Diffusion in International Relations

Norm diffusion occurs in contexts of interdependence between countries such as the current globalization process. It can be defined as a process that 'occurs when government policy decisions in a given country are systematically conditioned by prior policy choices made in other countries.' (Simmons, Dobbin, and Garrett 2006, 787). This broad definition allows for an analysis of the competition between the two normative frameworks on the liability of internet intermediaries outlined in the next section.

The literature identifies different types of norm diffusion that can be classified in four broad categories (Gilardi 2012). First, coercion is the imposition of a norm or policy by more powerful international organizations or countries, like in the case of conditionality associated with foreign aid. It was for example feared that the Transpacific Trade Partnership would impose regulations similar to US Digital Millennium Copyright Act to third countries (WikiLeaks 2014). Second, competition describes a process in which countries influence one another in their strategies to attract capital flows such as Foreign Direct Inversion. It has been argued since the late 1990s that the liability of internet intermediaries might harm the economic attraction of countries. Third, learning refers to the process through which a state evaluates the experience of other countries in order to calculate the likely consequences of a given norm or a given policy. Learning is certainly taking place in the diffusion of the two competing visions outlined in this chapter but the recent evolutions limit the ability of countries to fully understand the possible consequences of the adoption of norms such as the 'right to be forgotten'. Finally, emulation points to a diffusion primarily based on a logic of appropriateness regarding the normative and socially constructed characteristics of norms and policies rather than their objective consequences. Since the 'right to be forgotten' opposes two fundamental sets of rights (freedom of expression and freedom of information vs. privacy), emulation is certainly one of the main mechanism to be expected in the diffusion (or non-diffusion) of the norm.

The literature of norm diffusion focuses on the process rather than on the result of diffusion, which is useful to study a phenomenon of diffusion that is still under way. However, the norm diffusion literature tends to focus on the receiving end of the diffusion rather than on the diffuser. Receiving states are forced to, strategically chose to, rationally learn to, or are structurally encouraged to adopt a norm. In order to focus also on the diffusing end of the process and on the competition between two different normative frameworks, more traditional concepts related to diplomacy are required.

Exporting Internet Law

If we switch our attention to the diffusing end of norm diffusion, we should expect some active diplomacy in order to promote local norms as part of a foreign policy, especially in a context of competing visions. As developed in the following section, the different understandings that exist on both sides of the Atlantic regarding the liability of internet intermediaries do not only entail legal differences, they represent different visions on how to anticipate innovation in cyberspace and how to enforce existing (and new) rights on the internet. As such, both visions draw upon legal discourses on the right to privacy and on the freedom of expression. These legal discourses inspire publications, conferences, and debates in academic circles in different parts of the word while the courts' decisions serve as references both in academic debates and for further judicial decisions worldwide.

Socio-legal studies have long studied the transnational diffusion of legal norms. For example, Dezalay and Garth (2002) published a detailed account of the importation of legal concepts and norms from the US to Latin America through a mix of diplomacy and strategies of 'soft power'. These efforts helped establish a common language in order to identify and promote shared norms and values. Norm diffusion is not only a US project. In the case of the European Union, Léon Hendrickx, the first President of the main scholarly association for European Communities law (the Fédération Internationale pour le Droit Européen, founded in 1961), described it as 'the private army of European Communities' (Hendrickx, quoted in Vauchez 2015, p. 89). According to Vauchez, the association was 'furnishing, in colloquia and journals, the legal arsenal that would ensure the firepower needed for pan-European combat'. Latin America has historically been a recipient of both norm-diffusion diplomacies as illustrated by the importation of legal concepts such as the 'rule of law' (Rodríguez-Garavito 2012).

According to the literature, both the strategies to implement norm diffusion in the legal field, and the brokering/translating attitudes of receiving institutions might differ. First, in the case of the profound legal reforms implemented in Eastern and Central Europe after the fall of the communist regimes, the European Union and the US adopted different strategies to influence legal reforms (Hammerslev 2012). Whereas the US relied on private initiatives in order to export law through law firms, think tanks and legal assistance projects (though often funded by the US government), the European Union developed institutional projects aimed at states and representatives of state institutions such as judges and state officers. The US approach clearly fits within the scope of a public diplomacy with the creation of NGOs, scholarly networks and universities funded by US foundations. Public diplomacy in this case entails more than just traditional diplomacy; it is also a comprehensive strategy that includes private actors and that is geared towards the general public. The European approach remains closer to traditional diplomacy among state institutions. This difference is interesting since it seems to also apply to some extent to the case study of this chapter. There is a tendency in the US to focus on private initiative to replace an allegedly less efficient traditional state-to-state diplomacy.

The second variation is the diverging attitudes adopted by receiving institutions. Dezalay and Garth (2002) insist on the fact that the receiving end of law as an exportation product is not passive. For example, law firms, universities, and other 'cosmopolitan importers' tend to accept the legitimacy of specialized knowledge produced in the US and often adopt the vocabulary of the latest innovations by the legal field in the US. On the other hand, courts are less eager to use a cosmopolitan discourse and tend to rely on national logic and to respond to local contexts. Once again, this observation seems to be useful in the description of the diffusion of the 'right to be forgotten', where private initiatives from the US target NGOs and academia, while legal references by the Latin American courts seem to rely more on local legal traditions and jurisprudence, as well as to mentions to European rulings such as the Costeja case.

LIABILITY OF INTERMEDIARIES AND THE RIGHT TO BE FORGOTTEN: A TRANSATLANTIC DIVIDE?

The regulation of internet intermediaries by states and their liability regarding the content they host or distribute is a long-standing debate in internet governance. More than other internet governance issue, the liability of internet intermediaries illustrates the different perspectives that exist on both sides of the Atlantic.

The Liability of Internet Intermediaries and Internet Governance

Internet intermediaries supply the infrastructure that makes the internet work. As a category, intermediaries include a wide variety of entities such as Internet Service Providers, data hosts, websites, platforms, social networks, search engines among others (Riordan 2016, p.1). From a legal point of view, the situation is complex and evolves rapidly. Since intermediaries play a central role on the internet, it is one of the bottlenecks where online activities can be regulated (Zittrain 2006). However, most of the time the content displayed, indexed, and circulated by intermediaries is not created by them. Therefore, a legal debate emerged on the liability of intermediaries understood as a way to enforce fundamental rights online or as a threat to freedom of expression.

The legal issue of internet intermediaries has been discussed since the popularization and the commercialization of the network in the 1990s. Riordan describes four overlapping phases of international attitude. First, in the 1990s, during a phase of *disintermediation*, 'cyberspace' was considered an anarchical space (immunity of intermediaries in the most developed internet market: the US). Second, in the early 2000s, a phase of *protection* saw the creation of the first safe harbors that limited the liability of intermediaries (implementation of territorial sovereignty, but regulations such as the EU e-commerce directive or the US DMCA). Third, a phase of *expansion* of liability took place (late 2000s), in which new regulation were created both at a national and international level (e.g., copyright infringement regulations, international trade agreements). Finally, since the 2010s, internet regulation is undergoing a period of *balancing*, in which regulators seek to strike a balance between the liability of intermediaries and their fundamental rights. It is during this last phase that the European and the US perspectives became more differentiated. The difference in the treatment of intermediaries illustrate different visions of the autonomy of private actors, the role of the state and the balance between several fundamental rights such as freedom of expression and privacy. While both perspectives share the notion of 'limited liability' (MacKinnon et al. 2014), the understanding of the limitations differs as illustrated by recent developments on both sides of the Atlantic.

The European Perspective: Increased Liability of Internet Intermediaries

The current trend in the European Union is to differ from a previous consensus on the protection of intermediaries from liability. The e-Commerce directive adopted in 2000 by the European Union illustrated a shared vision according to which holding internet intermediaries accountable for content created by third parties was harmful for the development of e-commerce and might constitute a threat to freedom of expression. Liability was envisaged in terms of notice-and-takedown procedures when illegal or harmful content was hosted, listed, or circulated by internet intermediaries. The privileged approach was self-regulation and the directive called for the 'development of rapid and reliable procedures for removing and disabling access to illegal information; such mechanisms could be developed on the basis of voluntary agreements between all parties' (EU 2000, preamble, recital 40).

However, since the 2010s and during the 'balancing' phase of the regula-tion of intermediaries, the EU, European courts, and member states have taken a more pro-active stance in the regulation of the activities of intermediaries.

Two related historical evolutions explain the current trend in Europe. First, internet intermediaries that were once a myriad of small businesses have undergone a swift process of concentration. A handful of large plat-forms now control a large share of the markets of intermediaries such as hosting, web searching, social networking, and e-commerce among oth-ers. Intermediaries are not the vulnerable innovators anymore, in many cases they have become the incumbents they once unseated (Riordan 2016, 12–13). This makes regulation directed to intermediaries easier and more efficient. The second evolution is the identification of the threat for freedom of expression. In the late 1990s and early 2000s, any liability of internet intermediaries was seen as potentially undermining freedom of expression. Lately, the power of some intermediaries has shifted the attention towards the risk of private censorship induced by the self-regulation of internet giants (Kuczerawy 2015).

The result of this new trend is a case-by-case evaluation of the 'neutral' and 'purely technical' character of the activity of internet intermediaries. While in some cases, the lack of liability of internet intermediaries derived from the e-commerce directive still applies as the role of intermediaries is considered purely technical and not implying any treatment of the data[2], other cases illustrate the new European approach with a broader definition of what constitutes data treatment and thus a larger legal responsibility of inter-mediaries. For example, the European Court of Human Rights' judgment in Delfi vs. Estonia of 2015 appears in contradiction with former jurisprudence (Brunner 2016). Delfi, the largest Estonian internet news portal, was held responsible for user-generated content that appeared on its website in the form of comments, of which the portal lacked knowledge. The longer-term consequences of the Delfi decision might be important if online platforms are considered responsible for the content that is published alongside news articles. This issue is specially urging in a context marked by the legal and political challenges associated with fake news, as illustrated by the debate on Facebook's responsibility in the 2016 US presidential election.

The most important case for the present discussion is obviously the Costeja case. This case introduced a certain type of liability, whose diffusion in Latin America will be discussed in the following section. The Court of Justice of the European Union ruled in favor of Mr. Costeja in May 2014 regarding his claim that Google remove search results associating his name with an old news story about unpaid debts dating back to 1998. The Court established that Google was, in this case, a data controller:

the activity of a search engine consisting in finding information published or placed on the internet by third parties, indexing it automatically, storing it temporarily and finally, making it available to internet users according to a particular order of preference must be classified as <processing of personal data> . . . and the operator of the search engine must be regarded as the 'controller' in respect of that processing. . . . (CJEU 2014)

The delinking applies to items that are 'inadequate, irrelevant or no longer relevant, or excessive in relation to the purposes for which they were processes and in the light of the time that has elapsed'. This ruling epitomizes the European vision to hold intermediaries responsible for the data they treat, offering new leverage for a hybrid regulation based on public initiatives and private implementation (Chenou and Radu 2017).

The US Perspective: Continued Immunity for Internet Intermediaries

The US perspective on the liability of internet intermediaries has been more coherent over time. Basically, internet intermediaries are protected from liability regarding content created and posted by third parties. This view stems from a robust understanding of the First Amendment to the United States Constitution that protects free speech. Government intervention in the realm of free speech is exceptional and strongly disfavored. As a result, any court decision that would impose the removal of any user-generated content would be hindered by the provisions of the First Amendment (Holland et al. 2015). Additionally, as the internet became a promising vector of economic growth, the protection of intermediaries from liability was believed to foster innovation and to prevent flourishing internet businesses from the burden of legal responsibilities and regulations. Two main regulations set the foundations of the immunity of internet intermediaries:

First, one of the main sources on the immunity of internet intermediaries remains Section 230(c) of the Communications Decency Act more than twenty years after its adoption in 1996. As outlined earlier, while free speech is the main value that led to such regulation, the act sought 'to promote the continued development of the Internet and other interactive computer services and other interactive media [and] to preserve the vibrant and competitive free market that presently exists for the Internet and other interactive computer services'. The section 230(c) provides the basis for intermediary immunity: 'No provider or user of an interactive computer service shall be treated as the publisher or speaker of any information provided by another information content provider'. Section 230(c) of the Communication Decency Act covers claims of defamation, invasion of privacy, tortious interference, civil liability for criminal law violations, and general negligence claims based on third-party content but excludes intellectual property law, which are addressed by the US DMCA (Holland et al. 2015).

Second, the issue of intellectual property law, the Digital Millennium Copyright Act in its Section 512 establishes a safe harbor for internet intermediaries. Internet intermediaries are not responsible for copyright infringement by third parties under this safe harbor if they meet a number of conditions such as the lack of knowledge of the infringement, the lack of financial benefit from the infringement, and the fact that 'upon notification of claimed infringement as described in paragraph [the intermediary] responds expeditiously to remove, or disable access to, the material that is claimed to be infringing or to be the subject of infringing activity'. The creation of a notice-and-takedown mechanism avoided the issue of case-by-case evaluation and provided a simple mechanism in order to establish the immunity of internet intermediaries.

THE RIGHT TO BE FORGOTTEN IN LATIN AMERICA: AN OVERVIEW OF THE CURRENT SITUATION

The two visions outlined in the previous section propose two solutions based on different legal principles on how to regulate content in cyberspace. Both stress the need for legal norms in order to enforce fundamental rights while preserving the technical characteristics of an open and distributed network. The same question arises for Latin American countries facing the challenge of regulating cyberspace. While the situation is diverse in Latin America regarding the 'right to be forgotten', the analysis of the four largest countries of the region provides an overview of the general evolution of the debates, and of the influence of European and US norm diffusion[3]. As outlined in the following paragraphs, there is a clear trend towards moving away from a European-like perspective towards a rejection of the right to be forgotten and a full protection of intermediaries from liability.

Argentina

Argentina was a pioneer in the discussion about some sort of 'right to be forgotten' on the internet. Starting in 2006, several models and celebrities fought legal battles against Google and Yahoo Argentina that resulted, in some cases, in the de-listing of related content. In a sense, these rulings were precursors of the Costeja case in Europe. It is notable that the announcements informing a person trying to find one of the early de-listed pages in Argentina, are very similar to the announcement one finds on European Google domains when searching for Costeja's unpaid debt.

But since then, the results of actions asking to de-list personal information in Argentina seem to privilege the US vision against the liability of Internet intermediaries, as advocated among others by the Inter American System of Human Rights (Vargas 2016). In the famous first case that made Argentine a pioneer in the 'right to be forgotten' debates, the Argentinian Supreme Court finally ruled against the model Belén Rodríguez who sued Google and demanded that sites linking her name to pornographic content were deleted. This final ruling came after several decisions of lower instances recommending the implementation of a 'right to be forgotten'. Interestingly enough, the Supreme Court quotes the Costeja ruling to explain the importance of search engines for Internet access, but does not share the TJEU conclusions on search engines being data controllers (Corte Suprema de Justicia de la Nación 2014, § 10). The Supreme Court ruled against the liability of intermediaries except in cases where they do not take action in spite of previous knowledge that the content is harmful (and in exceptional cases of irrefutable illegality of content such as child pornography or direct and public incitement to genocide) (Corte Suprema de Justicia de la Nación 2014, § 21).

Nevertheless, there was no specific legislation about the liability of intermediaries in Argentina. In the aftermath of the Costeja case some bills were presented in both chambers, but like the judicial processes, they grew apart from a European vision of the liability of intermediaries and the 'right to be forgotten' towards a US vision[4]. The shift in the vision developed both by courts and by members of the Congress in Argentina illustrates the effects of an academic discussion on the 'right to be forgotten' and on the liability of intermediaries that clearly favored the US vision over the European one.

Brazil

The legal framework in Brazil is based on the Marco Civil da Internet adopted in 2014. Its article 19 provides a safe harbor for internet intermediaries unless 'after specific court order, they do not make arrangements to, in the scope and technical limits of their service and within the indicated time, make unavailable the content identified as infringing, otherwise subject to the applicable legal provisions' (Presidência da República 2014). Over the last few years, a number of bills has been proposed to complement the judiciary process with a 'right to be forgotten' more similar to what exists in the European Union (Artigo 19 2016). For example, the bill PL 7881/2014 makes it mandatory to 'remove links from internet search engines that refer to irrelevant or outdated data about the claimant'. Another bill (PL 1676/2015) also foresaw the de-linking of the name, image, or other personal aspects from facts, even true, that do not, or not anymore, represent a public interest. Both bills are currently under revision by the chamber of deputies and were merged in July 2016. However, given the current political situation in Brazil, the bills are likely to be removed from the political agenda.

At the same time, legal procedures under the Marco civil continued to be filed. While an estimate of one third of the complaints leads to content or link removal (Correio Barziliense 2016), there is no systematic and coherent jurisprudence. However, recent legal developments undermine the political will illustrated by the bills. Over the last two years, three court decisions have threatened the idea of a 'right to be forgotten' in Brazil. First, legendary TV presenter 'Xuxa' sought to remove all search results from Google that associated her name to criminal practices such as 'Xuxa pedófila' because of her appearance in an erotic movie from the 1980s, where she had sex with a twelve-year-old boy. The Superior Court of Justice ruled in favor of Google on the argument that search engines should not be forced to remove results stemming from a query on a specific term or expression or pointing to a specific text or photo (Folha de S. Paulo 2015). Second, another case (SMS vs. Google Brazil) regarding the removal of links pointing to nude photos confirmed the decision on the Xuxa case. In the ruling published in November 2016, the Superior Court of Justice further explained the lack of liability of search engines and repeated the argument of the 'Xuxa' case. Moreover, it insisted on the specificity of search engines as a category of service providers that cannot be held responsible for the content of the search results, since they cannot be forced to exercise *ex ante* control for all search queries (Superior Tribunal de Justiça 2016). Finally, a last case is still pending before the Federal Supreme Court concerning an old criminal case (Correio Barziliense 2016). The general attorney declared himself against a recognition of the 'right to be forgotten' by a court when it still does not exist in the Brazilian law. According to him, this would amount to 'limit the fundamental right to freedom of expression through censorship or through an imposition of previous authorization' (Canário 2016). Just like in the case of Argentina, a previous tendency to redraw the boundaries of intermediary liability towards an enhanced responsibility, and to implement a 'right to be forgotten' seem at least stalled.

Mexico

The most debated case in Mexico is similar to the Costeja case in the European Union. The entrepreneur Carlos Sánchez de la Peña asked Google to remove search results associated with his name, targeting a link to an article published by the magazine *Fortuna* on a corruption scandal (Pérez 2007). Google refused to remove the link. Sánchez de la Peña filed a complaint to the Mexican data protection authority (INAI, formerly IFAI), that ruled against Google (IFAI 2015) based on a 2010 privacy law (Ley Federal 2010). In the decision, the INAI directly mentioned the 'right to be forgotten'. The removal of the link—and not the taking down of the original article— resembles the 2014 decision of the Court of Justice of the European Union.

The decision triggered a heated debate, with NGOs such as Article 19 Mexico denouncing an act of 'censorship'. The fact that the plaintiff was a rich entrepreneur that allegedly took part in a corruption scandal, and the fact that the author of the article was a famous investigative journalist helped the mobilization against the INAI decision. The general tone of the news coverage, fueled by the declaration of civil society representatives, was clearly critical of the decision. An NGO, Red en Defensa de los Derechos Digitales (R3D), became the most vocal opponent and represented *Fortuna* in the appeal of the decision. According to the founder of R3D, the 'right to be forgotten' 'is harmful, it is a violation of freedom of expression. It is not possible to copy the judgment of the Court of Justice and bring it to the Mexican judicial system because it is not compatible' (Soto Galindo 2016). In August 2016, a Mexican court threw the INAI decision out (R3D 2016).

The cancellation of the INAI decision was celebrated as a victory of the freedom of expression by opponents of the 'right to be forgotten' (R3D 2016). In fact, the court only canceled the decision of the INAI based on the fact that they excluded *Fortuna* from the process, which meant that another process could have been initiated that should have included the magazine, as recalled by an INAI commissioner (Ureste 2016). However, the original plaintiff renounced to start the process over. In any case, it seems to represent an important precedent for other decisions based on the 2010 privacy law. Pending cases, such as the Búho Legal case, might follow up on the debate and even change the dominant mindset on the right to be forgotten in Mexico but the *Fortuna* case is clearly an important drawback for a potential application of the right to be forgotten in the country.

Colombia

Colombia is arguably the first country where a 'right to be forgotten' was explicitly mentioned in a court. Based on the 1991 constitution and the *habeas data* provisions, a magistrate evoked in a 1992 case a 'right to be forgotten' held by people in a world where databases do not forget[5]. However, following a Constitutional Court ruling from 2015 (Corte Constitucional 2015), there is not a 'right to be forgotten' on the internet in Colombia.

There is nonetheless an obligation for the original publisher to update, erase and take technical actions for making unavailable on search results false or outdated content relating to a person's intimacy, honor, and good name. This decision solved the Gloria v. Casa Editorial El Tiempo case without engaging the liability of internet intermediaries.

Indeed, Gloria argued that her right to a good name and her privacy had been violated in an article that mentioned her involvement in human trafficking, although she was not convicted following a prescription of terms. The court estimated that there was a collision between two sets of constitutional rights; on one hand, the editorial house's freedom of speech and society's right to access information, and on the other hand, the claimant's right to honor and good name.

During the proceedings, Google Colombia Ltda. defended a US perspective on the primacy of freedom of expression and declined any responsibility for information produced by third parties. The Court followed Google suggestion according to which 'content owners can decide which part of their content can be accessed by search engines and the newspaper was ordered to use technical measures, such as a robots.txt file, to ensure that search engines would not list the page referring to the prescribed crime.

Therefore, in Colombia, search engines and other intermediaries are relieved from any liability regarding content produced by third parties. Following the US perspective, search engines are considered mere intermediaries deserving protection on freedom of expression and right to information.

EXPLAINING THE SHIFT: US DIPLOMACY AND LEGAL DEBATES IN LATIN AMERICAN

In all four largest Latin American countries, two phases can be observed. Faced with the same regulation issues evidenced in Europe and based on similar values and legal systems, Latin American countries first considered extending the liability of intermediaries in order to protect the fundamental rights of users online. All of them discussed the implementation of a 'right to be forgotten' on the internet. However, after some lower instance judicial decisions and after the formulation of draft laws, the 'right to be forgotten' suffered drawbacks and was even abandoned as a concept in some of these countries. This shift represents a success for US diplomacy since it was able to export its visions of fundamental rights based on a technical a juridical discourse. At the same time, the European Union failed to gain leverage in global internet governance debates since its vision on the regulation of the internet was considered but mostly rejected by Latin American countries. In spite of European voluntary discourses following the Snowden revelations, EU institutions failed to 'defend [their common vision] jointly in the forthcoming international debates' (EC 2014).

The successful diffusion channels of the US vision are threefold: the inter-American intergovernmental network, the academia-civil society nexus, and the direct participation of US actors (especially internet companies) in political debates.

First, the Inter-American Human Rights System, and particularly, the Special Rapporteurship for Freedom of Expression, has been a major opponent of the liability of intermediaries and of the 'right to be forgotten'. The 2013 report on 'Freedom of Expression and the Internet' (IACHR 2013) illustrates the influence of the US vision:

> intermediaries should not be required to monitor user-generated content, and they emphasized the need to protect intermediaries *from any liability*, provided that they do not intervene specifically in the content or refuse to comply with a court order for its removal. (IACHR 2013, § 80, p. 35, emphasis added by the author)

This report is one of the main sources for Latin American opponents to the liability of internet intermediaries and to the 'right to be forgotten'. Since it was written by important law professors from the region, its academic influence is crucial in the development of the scholarly debates.

Second, the nexus between academia and civil society is an important part of diplomacy. US norms and values are diffused among civil society organizations and scholars through different channels. Given the dominant position of US scholarship, especially in internet law, research stemming from the US is influential in scholarly debates in Latin America. Citations and participation of US scholars in Latin American academic conferences evidence the strong presence of US ideas. For example, in the Argentinian case, the Centro de Estudios en Libertad de Expresión y Acceso a la Información (CELE) of the faculty of law at Palermo University has been an important participant to the debates with several publications on the issue (for example Ferrari and Schnidrig 2015; Vargas 2016). The publications are hostile to the 'right to be forgotten', describing it as a 'bad solution' (Ferrari and Schnidrig 2015, p. 10). They mostly quote US scholarship on the issue, especially from the Berkman-Klein Center at the Harvard Law School and the Stanford Center for Internet and Society, two major exporters of the US vision through norm diffusion.

Moreover, academic conferences have ensured an over-representation of the US vision in the region. For example, the event 'Libertad de expresión y derecho al ovoid digital' held at the University of Los Andes in Bogotá in

May 2017 featured several speakers from Harvard's Berkman-Klein Center for Internet and Society and Stanford's Center for Internet and Society as well as Google's chief lawyer for the region, all against the very idea of intermediary liability. This type of conferences does not only gather scholars but also policymakers and magistrates. As such, they are important conveyors of diplomacy. Often, speakers in academic conferences equate any form of intermediary liability to censorship. For example, Stanford's initiative on intermediary liability presents 'Liability regimes that put platform companies at legal risk for users' online activity [as] a form of censorship-by-proxy, and thereby imperil both free expression and innovation' (CIS n.d.).

The same dynamic occurs within civil society networks. Argentina's CELE as well as other civil society organizations and academic research centers participating in Latin American debates are signatories of the Manila Principles on Intermediary Liability, an initiative led by the US-based Electronic Frontier Foundation. Other examples include the Karisma foundation in Colombia and the global organization Article 19 that was an important actor in the debates in Mexico and Brazil. The first principle of the Manila Declaration echoes the US vision, as it reads: 'Intermediaries should be shielded from liability for third-party content' (Manila Principles 2015).

The fact that US scholars and civil society organizations are active supporters of the immunity of internet intermediaries does not stem only from the fact that they share a similar worldview with US internet giants and US political elites. The influence of internet companies and particularly of Google is much more direct and material. A quick look at the financial reports of major Latin American civil society organizations involved in the debates on the right to be forgotten shows that most of them received at some point funding from Google. While this does not necessarily imply any pressure from the company to impose its vision, recent reports show that Google has an active funding policy towards research that is supportive of the company's positions. According to the Campaign for Accountability's Google Transparency Project, many of the speakers in key academic and US-government conferences on privacy in 2016 had financial ties with Google (Campaign for Accountability 2016). The 2017 report on 'Google Academics Inc.' lists 331 papers published between 2005 and 2017 on public policy matters of interest to Google that received direct or indirect funding from the company (Campaign for Accountability 2017). The fact that a critic of Google's dominant position in digital markets has recently been fired from the Google-funded think tank New America Foundation illustrates that funding is expected to finance supportive research (Vogel 2017).

Third, internet regulations tend to implement a multistakeholder model that allows for the direct participation of the internet giants possibly affected by a 'right to be forgotten' and other liability regulations in the legislative process. As we have seen, Google Colombia was invited by the Court and its proposed solution was finally adopted. In the case of Argentina, in August 2016, the government created a multistakeholder working group on the internet, which brings together the national data protection authority and public policy representatives of US internet giants such as Facebook and Google (Listek 2016).

CONCLUSIONS: EU AND US INFLUENCE
IN LATIN AMERICA DEBATES

The case of the diplomacy of internet law shows the influence that can be exercised on the regulation of internet as a technical network by framing legal issues according to some fundamental values and worldviews. The US perspective in internet law is based upon the primacy of freedom of expression over other fundamental rights and the limited role of the state in the regulation of the network in order to promote private sector-led innovation. The European perspective foresees a greater role for the state and increasingly stresses the importance of privacy in the digital age, which leads to an extended liability of internet intermediaries. The two perspectives grew apart after the Snowden revelations. Although Latin America shares fundamental values with Europe and experiences the same challenges in terms of regulations of US internet giants, the European perspective faded away after some initial successes. This chapter has argued that this situation is partly the consequence of an active diplomacy by US actors in internet law. As in previous historical cases, the US perspective has been mainly exported by private actors, the academia and civil society organizations rather than by official institutions. Diplomatic efforts were able to stall the growing interest in Latin America in the 'right to be forgotten'. On the other hand, European institutions relied on the normative power of European decisions and failed to actively engage in Latin American debates. As a result, the European perspective lost strength in the region, which arguably undermines European ambitions to redraw the map of internet governance and to defend its common vision in international debates. Instead of establishing a strong dialogue between Europe and Latin America, or even to build a negotiation bloc to offer an alternative to the dominant power in the digital age, the European Union became more isolated in its efforts to regulate cyberspace.

NOTES

1. The author would like to thank Natalia Carrizosa for her help on the mapping of the debates on the 'right to be forgotten' in Latin America and for her helpful comments during the drafting process.

2. For example, Google vs. Louis Vuitton and others, ECJ, C-236/08 to C-238/08, and Scarlet Extended SA, ECJ, C-70/10. For a discussion, see (Männiko 2014).

3. Smaller countries such as Uruguay and Costa Rica have a different take on the 'right to be forgotten' and their analysis is important as well. However, the present paper focuses on the largest countries.

4. After a 2010 bill foreseeing the liability of internet intermediaries, Senator Pinedo proposed a new bill (S-942/16) in 2016 with a more moderate approach. In October 2016, his bill was merged with Senator Fellner's bill (S-1865/15) that explicitly protect search engines from liability. See http://www.senado.gov.ar/parlamentario/parlamentaria/verExp/parla/S-1865.15-PL#, last accessed August 30, 2017.

5. I am thankful to Prof. Nelson Remolina Angarita for pointing me to the Decision No. T-414/92 of the Colombian Constitutional Court as one of the origins of the 'right to be forgotten'. See https://habeasdatacolombia.uniandes.edu.co/?page_id=15.

Chapter Eleven

Policy Diffusion and Internet Governance

Reflections on Copyright and Privacy

Krisztina Rozgonyi and Katharine Sarikakis

The Internet is a site of struggle for control over infrastructure, content, and users. The following discussion interrogates the struggles over the governance of privacy and copyright, as two particular sites that have become increasingly associated with factors affecting the freedom to access and use content on the internet, as well as expression and political participation. An integral part of governance is the process of policymaking and the role of principles, actors, and structural factors within it. As states and global actors increasingly seek new models of legislation and regulatory formats, and as global communication presents societies with challenges not always addressed at local levels, unavoidably both the formulation of policy 'problems' and their 'solutions' become global 'products' in this exchange. Perhaps now more than ever before, the search for common policies, underpinned by commercial interests as well as social needs for global coordination, provides a fertile ground for the study of policy 'flows', that is, policy models that are copied, amended, adjusted, repeated in legislative efforts across countries. Moreover, due to the combination of territorial, financial, and technological integration processes, policy areas lose part of their distinct responsibilities and become enmeshed in relations of interdependence. It has become increasingly evident that the hitherto protection of existing rights and the implementation of emerging policies and policy regimes clash, as the implementation of some policies areas depend on the infringement or compromise in the protection of fundamental rights, such as is the relation between copyright and privacy or security and privacy (Sarikakis and Rodriguez-Amat 2013). The predominant catalyst in this process is the fact that internet-related policies turn towards the regulation and policing of individuals and individual behaviour rather than towards the conduct and operation of corporations or other market and policy-affecting actors. This emphasis on individuals inevitably causes abrasions to the solidity of citizens' rights, with the most worrying dimension that observed in human rights and fundamental freedoms.

In this chapter, we deploy the concept of policy diffusion to study the 'processes through which the framework of political, institutional, and cognitive actions is used or readjusted for the development of action frameworks within other political systems, often foreign' (Bevir 2006, 705) and draw on work examining the dynamics of policy diffusion in international and intergovernmental constellations (Sarikakis 2004; Chakravartty & Sarikakis 2006; Sarikakis 2012b; Sarikakis and Ganter 2014). We examine whether the distinct policy issues of informed consent as part of privacy policy and graduated response as a way to enforce copyright provisions, present the characteristics that indicate policy transfer and between which directions. The selected areas constitute cases of governance that affect citizens 'everyday life by interfering with citizens' civil rights and fundamental freedoms to informational freedom and right to privacy. These sites of struggle, we argue, entail policy principles that reflect ideological dispositions about the role of the state and the market, the role of the citizen as an actor and ultimately even the state of fundamental rights in a volatile world. For this reason, we trace the process of policy across various stages and discuss intersections of continuity and transformation. The two sites of struggle represent two, seemingly, diametrically opposite ends of the policy aim spectrum: graduated response is a punitive measure, aiming to deter individuals from a specific act while using the internet and accessing content. Informed consent aims to provide for a positive right for the individual when attempting to use online services, including accessing and generating content. Both measures assume a third party involved in the relation of the individual to content, and consequently determining the regulatory direction, that of private market actors. Although in both cases it is the individual placed as the addressee of policies, and on whose shoulders the weight of implementation falls, it is ultimately the private market entities that maintain their power position: in both cases, corporations can de facto protect their 'ownership' more effectively than individuals, by withdrawing content and/or services at a disproportionate degree of 'loss' vis-à-vis the individual.

'INFORMED CONSENT' AS EU PRIVACY POLICY STANDARD

Public concern about the role of privacy policy within emergent global information policy regimes is not novel post-9/11 but has grown since individuals came to personally experience privacy intrusions (Braman 2004, 37). Corporate entities and states have tapped zealously into the possibilities offered by mass accumulation of data on individuals, generated and collected through online and digital communications globally. The post-Snowden era is clearly characterized by the struggle to gain increased control over personal data and their processing. The right to privacy, including the protection of individuals' data is a fundamental right, however, its safeguarding in the era of big data and algorithmic decision-making comes under immense pressure (CoE 2013).

Within regulatory frameworks, the principle of 'consent' given by the individual in regard to making available of their personal data while using services online has been linked to fundamental rights on 'informational self-determination' (Recht auf 'informationelle Selbstbestimmung'[1]) by the German Constitutional Court (Bundesverfassungsgericht) since its landmark decision in 1983[2] ('Volkszählungsurteil'). The court's decision introduced a new jurisprudential ideology of the contours of personality, amid growing concerns about threats to privacy posed by the use of electronic data processing. The far-reaching impact of the decision also served for the first policy transfer cases in Europe (e.g., in Austria, Hungary) reflecting 'copying' characteristics and embedding privacy policy in fundamental freedoms.

Regulation on Privacy and Electronic Communications 'Consenting' represents the expression of a right and power for self-determination and the integrity of personhood and is particularly important as a measure of bringing back a 'lost' equilibrium of power, particularly in situations where such imbalance presents itself de facto: individual vis-à-vis the state, corporations, and other powerful individuals. Historically and jurisprudentially, 'consent' transverses the dichotomy of public vs. private on one hand, whereby private sphere actions among two or more actors still require expressed consent by all parties.

The question on what constitutes public vs. private policy terrains within emerging global information policy regimes as contested features thereof were also strongly connected to debates on privacy. Informational self-determination-based data protection regimes were the subject of scholarly and policy critiques, on the basis of the asymmetry of power between individual data subjects (users) and of data processors in the privacy marketplace and the 'knowledge gap' as a systematic consequence of the social and institutional structure of personal data use (Schwartz 2002, 72). This asymmetry has favored those who had the authority to exercise control, while others routinely waived their rights without comprehensive understanding on the issues at stake, pre-empting the right itself (Mayer-Schönberger 1997). Policy failures, however, have led to the emergence of a new generation of data protection legislative and policy actions and the standard of the 'informed consent' as a renewed attempt to provide for individuals' authority on self-determination or to counterbalance the dominance of private entities. This standard and related policies were aimed to increase public trust in technologies and, hence, allow for the faster adoption of technologies and ecommerce; as a secondary goal they aimed to provide a remedy to what has been seen a threat to fundamental freedoms. Given this eminent and overriding impact of 'informed consent' we are first looking into its policy trajectory within the EU and then into diffusion elements beyond the boundaries of Europe.

Informed consent as a privacy policy standard was the subject of different data protection legislative and regulatory regimes. Historically, it was the European Convention on Human Rights (ECHR, CoE 1950) that first stipulated for the right for respect of private life (Article 8) at the European level. Although the EU as entity is not a party to the ECHR, all its member states (MSs) are individually parties, therefore bound by its rules. The protection provided focused mainly against interference by public authorities, an approach that needed reconsideration in the wake of information and communication technologies and the role private actors played in their deployment. The jurisprudence of the ECHR[3] has further elaborated on the right. The ECHR rulings set a requirement for the concept of 'consent of the individual' to third party access to their personal data to be introduced in national legislation.

The Council of Europe (CoE) laid down the minimum standards on processing personal data in the public and private sectors with the requirement for a 'fair and lawful way'[4]. 'Consent' was not specified as a requirement, then, however, later, the CoE addressed the scope of the standard by specifying the free, specific, and informed consent of the data subject as the key legal basis for profiling.[5]

The EU Charter of Fundamental Rights in primary legislation (EU 2012) recognizes 'consent' as the basis for lawful data processing (Article 8(2). In EU secondary legislation, the 1995 Data Protection Directive provides the highest level of legal recognition of 'informed consent' requiring that the data subject unambiguously give their consent to data processing (EU 1995, Article 7 (a)), (also) in the process of data transfer to a third country, by giving their consent unambiguously to the proposed transfer (EU 1995: Article 26 (a)). The interpretation of the scope and content of the standard requires case-by-case analyses, which is beyond the scope of this chapter. However, there are certain indicators—availability, accessibility, and visibility of the information and the language used—that were foreseen as important elements (European Union Agency for Fundamental Rights and Council of Europe 2014).

The ePrivacy Directive (EU 2002) set additional criteria on 'clear and comprehensive information' provided to the user as a prerequisite by 'electronic communications networks to store information or to gain access to information stored in the terminal equipment of a subscriber (EU 2002, Article 5(3)), referring to the use of cookies by relevant providers. Whilst 'informed consent' gained specific importance in behavioral advertising, the rules demanded further elaboration. Opinion 2/2010 of Article 29 Data Protection Working Party (Article 29 WP 2010) envisions the creation of prior opt-in mechanisms and affirmative action by data subjects to obtain their informed consent on receiving cookies or similar devices and the subsequent monitoring of their surfing behavior. Furthermore, it recommends application of time limits, revocation possibilities and visible signaling while monitoring takes place. Opinion 15/2011 of Article 29 Data Protection Working Party (Article 29 WP 2011) is until now the most detailed, yet legally not binding, policy document that has developed key elements of the consent standard, interpreted various options of consent-giving as 'indication', 'freely given', 'specific', 'unambiguous', 'explicit', 'informed' with the aim of providing guidelines to the validity of users' actions in this regard.

The parallel disruptions of technological and legislative advancements project a potential 'regime change' in European information privacy policies (Mayer-Schönberger and Padova 2016). Over a period of four years, the reform of EU data protection rules concluded in legislative acts in the spring 2016: the General Data Protection Regulation (GDPR, EU 2016) replaces the Article 29 Working Party by 25 May 2018 and includes strengthened consent as a basis for collecting and using individuals' personal data (EU 2016, Preamble (32), Article 4 (11)). 'Consent' is applicable in regard to flows of personal data to and from countries outside the EU (EU 2016, Article 44). In terms of data transfers based on consent, the GDPR requires explicit consent . . . 'to the proposed transfer, after having been informed of the possible risks of such transfers for the data subject' (EU 2016, Article 49 para. 1 (a)).

The trajectory of informed consent illustrates its central role within European privacy contestations. It is also to suggest that the EU was expected to represent its valued privacy standard beyond its legislative authoritative power in external relations whereby the protection of European citizens could have played a significant role. Protection of personal data has been perceived as a core European value in EU policymaking.[6] Since the 1995 DPD, there have been significant efforts in presenting EU privacy policies as globally present and model-setting. The EU was keen on enforcing its policies beyond EU borders as a counterbalancing act in regard to a level playing field for European businesses.

There have been claims that strict EU data protection rules actually challenged competitiveness of European market players vis-à-vis global competitors being subject to costly and burdensome administrative rules on data safety. To overcome trade-offs of data protection policies, the EU attempted to 'export' its policies and make others in trade relations with Europe to implement similar level safety standards. With the introduction of the 'adequacy of the level of protection' (EU 1995, Article 25 Principles) the EU has established a framework of assessment to and from third country data transfers. The adequacy assessment is the main governance framework of EU privacy policy transfer vis-a-vis the US and US-based corporations. Further research needs to explore the extent to which it has secured the set aims of EU policies in regard to safeguards of EU citizens' fundamental rights[7]. It can be argued, with the assessment process the EU put in place an effective tool enabling the evaluation of the level and appropriateness of actual transfer of its policies as well.

THE EU AT SUPRANATIONAL FORUM—THE OECD FRAMEWORK ON PROTECTION OF PRIVACY AND TRANSBORDER FLOWS OF PERSONAL DATA

EU member states and the EU are represented in a multinational cooperation set up by the OECD since the mid-1970s. The adoption of the OECD Privacy Guidelines in 1980 constituted the first internationally agreed upon set of privacy principles to regulate the flow of personal data across borders at an agreed level of data security. The Privacy Guidelines (OECD 1980) set the 'scene' for later policymaking by the CoE Convention No. 108 (CoE 1981) and for the adoption of the Data Protection Directive (EU 1995) by the EU. This also meant, historically, the EU's own privacy policies have first reflected and embedded the multinational agreement within the OECD framework and have taken it forward from there.

The OECD Privacy Guidelines (OECD 1980) set the privacy standard to data transfers at 'knowledge' (minimum requirement) or 'consent' of the data subject (OECD 1980, Para. 7). Also, they have allowed for options 'where appropriate' for derogations such as in cases of criminal investigation activities or routine updating of mailing lists. The scope of 'consent' was not specified any further. The OECD Privacy Guidelines were revised in 2013. The new set of rules (OECD 2013, para 7, 9, 10) are verbatim with regards to the 'consent' standard. There are a few possible interpretations to this (Marcinkowski 2013), but from a policy transfer perspective we can argue, that no change of policy within thirty-three years within this highly sensitive terrain and against the policy development results within the EU is rather indicative for failure of transfer.

The Safe Harbor Privacy Principles (US DoC 2000) were agreed upon after the DPD (EU 1995) became effective in 1998 to provide clarity on what constitutes an 'adequacy standard' according to EU law (EU 1995: Article 25). Qualification by US corporations for the safe harbor was entirely voluntary, and compliance enforcement was with US authorities. With regards to consent-driven privacy policies, they entailed a limited and lower-level protection to EU citizen data than the DPD (EU 1995). The general requirement was an opt-out standard, opposed to the opt-in regime of the DPD (EU 1995), and opt-in only in case of sensitive data. In terms of transfer characteristics, the Safe Harbor framework showed indicators of 'copying' and of limited similarity of policy models and of regulatory authorities (Sarikakis and Ganter 2014).

For more than a decade, the Safe Harbor agreement was in operation under these conditions. Despite 'widespread expectations' about the EU as 'a global actor with the ability to express coherent policy preferences internationally' and creating a 'global standard that reflects its values: the European Data Privacy regime' (Heisenberg and Fandel 2004: 109, 110), the Safe Harbor framework a case of reverse policy diffusion, that is of the lowest denominator originating from US jurisprudence. The period of 'smooth' cooperation between the EU and the US basically ended on 25 June 2013, when an Austrian citizen—a privacy activist—Maximillian Schrems filed a complaint to the Irish Data Protection Commissioner 'by which he in essence asked the latter to exercise his statutory powers by prohibiting Facebook Ireland from transferring his personal data to the United States' (CJEU 2015: 15).

Schrems challenged Facebook Irish subsidiaries, with regards to his personal data transferred to the US first in the Irish Data Protection Commissioner, who refused to investigate. Then Schrems made a request to the European Court of Justice for a preliminary ruling. The European Court of Justice rendered the previous 'Safe Harbor' framework invalid and compromising the essence of the fundamental right to respect for private life as well as to the fundamental right to effective judicial protection (CJEU 2015). The enactment of the EU-US Privacy Shield Privacy Shield Agreement on cross-border data transfers from the European Economic Area to the US (EC 2016) followed the ruling and transfer of privacy policies gained a new momentum since then.

The Schrems case as a form of citizens' activism on international policymaking and on supranational level policy transfer is to be viewed as a landmark action in regards to global policy regimes. It is a case of emerging activism beyond national borders with a significant impact to foreign policy.

Constructivist theories on international relations argue for social embedded-ness of power politics and for the complex construction of identities and interests of nation-states (Wendt 1992; 1999).

The EU-US Privacy Shield Agreement (EC 2016) and the adequacy deci-sion by the EC[8] was adopted in July 2016, with the aim of better enforcement of EU data protection policies and principles and enforcement of stronger obligations on companies in the US to protect the personal data of individu-als. According to the new rules, US companies and individuals will have to register to the Privacy Shield list and self-certify that they meet data protec-tion standards set out by the arrangement. It is up to the US Department of Commerce to monitor and verify compliance with the standards.

The standards are set by the 'Privacy Principles' attached to the agree-ment (EC 2016, Annex 2). They are to serve for criteria based on which the verification of any corporate or individual application is provided, and constitute normative embedding of transfer of 'informed consent' privacy policies as well. Consent-based standards are applicable only to sensitive data and require affirmative express consent (opt in) if such information is to be disclosed to a third party or used for a purpose other than those for which it was originally collected with several eligible exceptions (EC 2016, Annex 2, II. Para. 2.c; III. Para. 1.a). From a policy diffusion perspective, it must be noted, that the scope and the breadth of the privacy standard are—again—limited compared to the EU standard level as set by the GDPR (EU 2016) not reflecting EU policy recommendations.

There are at the moment 1,828 US-based companies signed up for the Pri-vacy Shield List[9], including Facebook, Google, Microsoft, and Twitter. The role of these corporations, which effectively act as global information inter-mediaries, in implementing privacy principles necessitates further scrutiny into their practices through the lens of informed consent. Internet intermedi-aries have historically shown their privacy policies are 'obfuscate, enhance, and mitigate unethical data handling practices' (Pollach 2005, 221). Sargsyan argues '(t)hese private companies have mastered the strategy of implement-ing seemingly privacy-centric policies and tools without compromising their economic interests' (Sargsyan 2016, 10).

Twitter, with 313 million users worldwide and 79 percent of accounts outside the US,[10] and Facebook with more than 1.23 billion daily active users on average[11] constitute beyond any doubt the largest global communication platforms today. Their terms and conditions,[12,13] and increasingly copyright legislation govern the conditions of communication taking place on these platforms. The disconnect between users and intermediaries with regard to expectations of consent and privacy has a major impact on trust of users in social media (Custers, van der Hof, and Schermer 2014) and more important in media as a communicative space (Sarikakis 2015) where people enjoy fundamental rights as citizens.

GRADUATED RESPONSE

Graduated response or three-strikes policy is a copyright control scheme that deals with copyright infringement by individuals, as a response to the creative industries' concern with profit loss deriving from downloading and peer-to-peer file sharing (Ranaivoson and Lorrain 2012). Deriving from the baseball rule 'three strikes and you're out', the scheme was renamed graduated response to minimize its negative connotations of jail penalty (Haber 2011). Policymakers and the industry still use the term three strikes, demonstrates the action ISPs can take before the user is 'struck' for the last time (Yu 2010)[14]. IPs are being delegated the role of Internet police, as it is them who have to identify the infringers and trigger the graduated response process. It is argued that culture was used as an excuse to protect the cultural industries with arguments over the clashing role of the Internet as a tool for the industry and as a fundamental right for the citizens (Meyer 2012).

  With the rise of peer-to-peer technologies, global music sales and licensing decreased in half during the period 2000 to 2010 (Danaher et al. 2014). Graduated response was first implemented around 2008, in order to counter individuals' file sharing online. Some European countries such as the UK and France adopted graduated response in their legislation as did Taiwan, South Korea, and New Zealand (Giblin 2014), Switzerland[15] and Nigeria[16]; there are also initiatives in Ireland and the US that are similar to graduated response but take place on private rather than State level, therefore not as hard law. In the US in particular, three strikes has become six strikes as is the case of the US graduated response system.

  The direct impact on citizens' access to content resulting from copyright legislation led to the rise of protest in the form of social movements. Dobusch and Quack (2008) classify piracy parties as significant actors in the broader copyright movement of the young generation—the 'digital natives'. A unique example of social movement close to the struggle over copyright is the Pirate Party. In 2009 the Swedish 'Pirat Partiet' reached 7.1 percent of the votes and joined the European Parliament, an impressive result for a special interest group (Dobusch 2009). The movement-based party reached political success in other countries such as Germany and is present in various regions of Europe, Latin and South America or Pacific countries. All branches share the view that Internet related copyright regime should be combated and act against criminalization of private file sharing practices. The movement

Krisztina Rozgonyi and Katharine Sarikakis

promotes open source and free software and speaks for net neutrality. The response of governments has been dominated by a punitive logic by restriction to the internet and therefore content. There are two characteristics of importance here, common across the following selected cases: the privatization of policy implementation for use restrictions and the punitive measures over the individual. As we will see, there is an overall turn towards 'educational' and so preventive, rather than punitive measures, predominantly deriving from the ineffectiveness in challenging and changing behavior.

The first attempt of the British government to intervene on illegal file-sharing was the 2008 Creative Britain report that encouraged rights holders and IPs to collaborate in order to address the issue (Barron 2011) and the Digital Britain report and the subsequent Digital Economy Bill in 2009 (Mendis 2013). The Digital Economy Act (DEA) of 2010, which succeeded the Communication Act of 2003, includes a graduated response scheme that involves ISPs with more than 400,000 broadband lines. According to that, ISPs use standard notification forms for their customers if there is evidence of copyright infringement. The scheme includes an initial and an intermediate notification and the possibility of appeal for those allegedly breaking copyright law. DEA has been viewed to be shortsighted when it comes to the issues around those individuals who are infringing copyright, time-restricted, as well as disproportionate and unsustainable with reference to website blocking (Mendis 2013).

ISPs and rights holders have agreed a new framework on copyright infringement on the basis of educating those allegedly infringing the law. This includes four warnings, but the scheme is educational and does not include enabling instruments for legal action to be taken by copyright holders.[17]

As the creative sector in France is among the largest in the European area, the implementation of graduated response in France is not accidental (Meyer 2012). The French entertainment sector suffered a loss of €1.2 billion of profit as a result of digital piracy in 2007 only, a couple of years before the HADOPI law came into enforcement, with the French National Assembly reporting a job loss of about 5000. The video sector seems to have suffered the greatest losses, of €605 million, and the music industry €369 million and the book industry €147 million (Aroba 2013).

The first HADOPI law of 2009 was deemed unconstitutional (Suzor and Fitzgerald, 2011); the second, HADOPI-2, followed in 2010, created an authority to look after copyright infringement and identify ways of protecting intellectual property online. Penalties for infringers include suspension of Internet access for up to one year and a fine of up to €1500 (Giblin 2014). The French video games and book industries do not turn to the HADOPI authority to protect their interests, as opposed to the music and cinema (Lausson 2020).

The introduction of HADOPI caused great concern and was subject to protest by social groups including Anonymous, with disruptions of computer systems, contesting companies' performance or policies; they also have a voice in the discourse about freedom of speech and digital rights (Milan 2013). Street art and graffiti in public space, which was painted over by the disturbed local authorities, were also inspired by this (Vatu 2014).

In May 2013, the Lescure Report evaluated HADOPI as ineffective in achieving its aims, which although having reduced P2P infringement to an extent, it did not manage to turn infringers to legitimate markets. It recommended to abolish the HADOPI authority, reduce the infringement fee to €60 and abolishing termination of access to the Internet as a penalty (Giblin 2014). The result of this report was HADOPI-3, according to which Internet suspension remained as a penalty but only because it was not possible to abolish it with a simple decree (Giblin 2014).

Traces, or implications, of the graduated response can be found in the EU legislation, in particular in the Telecoms Reform Package aimed at creating a common regulating set for all member states (EC 2009). The Draft Package introduced two amendments related to copyright and in favor of graduated response (Suzor and Fitzgerald 2011).

Copyright gained attention in public debate in 2008 when the citizens' group La Quadrature du Net initiated a campaign against copyright enforcement in Europe through a mass email campaign sent to the European Parliament in reference to its first reading of the Telecoms Package Reform (La Quadrature du Net, 2009). Activists' attention was turned to the fact that there was evidence that a graduated response policy at the European level was going to be implemented after the revision of the Privacy and Electronic Communications Directive (Breindl and Briatte 2013). After intense public debate (Euractiv 2009; Boardman 2011) the Package finally included what is known as Amendment 138, or Freedom Provision, which had the approval of the Parliament, the Commission, and the Council (Ferti 2010). Citizens' action succeeded in a reminder of the obligation for the member states to respect the ECHR, included in the third reading in 2009 by the replacement of Amendment 138 (EP 2009).

The US case is an example of graduated response enforced not by State legislation, but as a result of private agreements. In the US, Internet piracy is estimated in losses in sales of more than $25 billion every year (Boardman 2011). The Copyright Alert System (CAS) or six strikes comes from a Memorandum of Understanding signed by ISPs, the Independent Film and Television Alliance (IFTA), the Motion Picture Association of America (MPAA), the American Association of Independent Music (A2IM) and the Recording Industry Association of America (RIAA), with the aim to notify users alleged to infringe copyright and educate them on the consequences of infringement (Owen 2012). This scheme uses six warnings, instead of three, and four steps, and as it is not combined with legal action there are no legal consequences and no Internet suspension foreseen, although the Internet speed may be reduced and other measures can be taken for infringers (Aroba 2013; Owen 2012).

SUCCESS, FAILURE, OR NEITHER?

Regarding the major legal deficiencies of the Safe Harbor framework that led to overturning by the Court of Justice of the European Union, from the policy transfer perspective the EU appears to have failed to enforce its privacy principles. Citizens challenging established policy led to the GDPR, the EU-US Privacy Shield Agreement (EC 2016) and the Legislative Proposal for Regulation on Privacy and Electronic Communications (EC 2017a). As previous experience on practices of private actors such as Twitter or Facebook suggests, even in the light of enhanced control mechanisms set by the Privacy Shield, implementation of EU privacy standards is still a work in progress. In this case the answer for policy failure or success lies with the next actions of the EU following and enforcing the values and principles incorporated in those texts, and eventually reconsider whether consent-based privacy policies were best to safeguard EU citizens' rights at the age of big data.

The OECD 2013 update of Privacy Guidelines case is an example of policy failure. The fact that the revised Guidelines are not reflecting any of the policy efforts the EU has made all through the years since 1995, with special regard to the Article 29 Working Party policy streams on consent-based privacy standards, is a significant marker of unsuccessful policy representation.

The policy transfer trajectory in the graduated response scheme appears to be implicit, with transfer taking place in a lobbying level and it seems there are still pressures from the creative industries to maintain it; the DEA in the UK and HADOPI-2 in France were introduced after the EU Telecoms reform, with the latter being revised in a HADOPI-3 in 2013.

In the EU, graduated response appears to be still implemented but with doubtful results, even in those countries that have a very active music and creative industries, such as France. The HADOPI case has been successful in identifying users involved in copyright infringement. Since 2010, more than 1000 cases have been transferred to justice; from June 2015 to June 2016 688 cases were brought to justice, as opposed to 362 from 2010 to 2015 (V. J. 2016). The HADOPI authority released the numbers until the end of January 2017, according to which from the 8,126,738 first notifications to users since

2010, 340 suits were known to HADOPI until the 31 January 2017, from which 165 ended up in alternative measures to be taken towards infringers and only 99 condemned by courts (HADOPI 2017).

Graduated response has been criticized for its impact on the right to freedom of expression and privacy (Giblin 2014) as well as being part of a new direction in copyright law that targets and intimidates citizens (Ranaivoson and Lorrain 2012; Haber 2011). It has been considered ineffective in all its proclaimed aims, i.e. minimizing copyright infringement, turning users to the legitimate market or assist copyright law to achieve its aims (Giblin 2014). There are also concerns expressed with reference to the degree of surveillance and the regulation of the users' internet activity (Meyer and Van Audenhove 2010) and on its impact on the fundamental right of freedom of speech (Haber 2011).

The relative success of transposing the intra-EU privacy regime internationally as forging a 'European identity' towards the US was argued to be made possible—from a constructivist perspective—as a significant effort aligning the interests of a broad number of stakeholders (Heisenberg and Fandel 2004). The impact of the Schrems case on resetting the privacy agenda between the EU and the US implies that foreign power politics is very much shaped today by new actors and by their values and interests, emphasizing the need for re-considering the relevance of constructivism to foreign policy analysis (Behravesh 2011).

In the case of graduated response, citizens also turned to supranational entities such as the European Parliament to stop a process of policy transfer that was breaching their rights and criminalizing users. The collective action instruments used in this case were mainly public awareness and citizen's mobilizations (Breindl and Briatte 2013). With these, citizens' movements succeeded in placing the issue in the agenda of the European Parliament and eventually stopping graduated response from becoming a policy instrument in the European Union.

The Internet Freedom Provision, resulting from the intense debates during the Telecoms Reform Package particularly from the European Parliament who wanted to introduce judicial oversight of any Internet disconnections, as a strict answer to the French HADOPI law, is the first document that relates access to Internet with citizens' fundamental rights and freedoms online (Ferti 2010). Although the Telecoms Reform Package would still allow graduated response schemes, the citizens outcry that stemmed from HADOPI and the debate on an EU-institutions level during the Package reform, demonstrate that without citizens' outcry nothing would have changed and make this case an example of success stemming from failure.

The consequences of informed consent may be indirect, as citizens do not know what happens after they 'click away' their rights, but in the case of graduated response they are immediate and harsh, which has triggered citizens' reaction and outcry. Social movements appear to be triggered more directly in cases of visible violation of citizens' rights, such as the graduated response, and their actions lead to improved awareness of the role of technology and its regulation to modern democracies. 'For the first time in history, these issues mobilize a broad and diverse public that also includes non-specialists' (Milan 2013, 6). Furthermore, due to the protests some related issues such as net neutrality and censorship entered the extensive public debate (Horten 2013).

European data protection policies have been developed to protect fundamental rights and values of EU citizenry, including the right to privacy. The question is not whether to apply them in the age of big data (EDPS 2015), but rather how to implement them in a meaningful way. There are also lessons to be learned regarding the trajectory of policy transfer initiated or 'suffered' by the EU. Further research should consider the key characteristics of these transfers and their degrees of failure or success.

NOTES

1. BVerfGE 15.12.1983 65,1.
2. BVerfG, 15.12.1983—1 BvR 209/83; 1 BvR 269/83; 1 BvR 362/83; 1 BvR 420/83; 1 BvR 440/83; 1 BvR 484/83.
3. L.H. v. Latvia (no. 52019/07), Perry v. the United Kingdom (no. *63737/00*), L.L. v. France (no. 7508/02), Peck v. the United Kingdom (no. 44647/98), Gaskin v. the United Kingdom (no. 10454/83), Odièvre v. France (no. 42326/98), Godelli v. Italy (no 33783/09).
4. Articles 1 and 5 of Convention 108 (CoE 1981).
5. See (CoE 2010), III.3. p.119.
6. 'Being European means the right to have your personal data protected by strong, European laws. . . . Because in Europe, privacy matters. This is a question of human dignity.' (Juncker 2016).
7. See the list of companies for which the EU BCR cooperation procedure is closed; accessed from http://ec.europa.eu/justice/data-protection/international -transfers/binding-corporate-rules/bcr_cooperation/index_en.htm.
8. COMMISSION IMPLEMENTING DECISION of 12.7.2016 pursuant to Directive 95/46/EC of the European Parliament and of the Council on the adequacy of the protection provided by the EU-US Privacy Shield.
9. Privacy Shield List; accessed from https://www.privacyshield.gov/list.
10. Twitter Usage/ Company Facts available at https://about.twitter.com/en/who -we-are/our-company.
11. Facebook Company Information available at http://newsroom.fb.com/com pany-info/.

12. On Twitter Privacy Policy see Information Sharing and Disclosure rules available at https://twitter.com/privacy?lang=en.

13. Facebook Data Policy available at https://www.facebook.com/policy.php.

14. For three strikes vs. graduated response terms see Yu, 2010.

15. See http://graduatedresponse.org/new/?page_id=621.

16. For a copy of the Copyright Bill see: http://graduatedresponse.org/new/wp-content/uploads/2016/02/DRAFT_COPYRIGHT_BILL_NOVEMBER-_2015.pdf.

17. Source: http://www.bbc.com/news/technology-38583357.

List of Acronyms

4S	Society for the Social Studies of Science
A2IM	American Association of Independent Music
AAAS	American Association for the Advancement of Science
ACNU	Asociacion Cubana de las Naciones Unidas
ACTA	Anti-Counterfeiting Trade Agreement
AfICTa	Africa ICT Alliance
AfriNIC	African Network Information Centre
ALAC	(ICANN) At-Large Advisory Committee
APC	Association for Progressive Communication
APIG	Association for Proper Internet Governance
APNIC	Asia-Pacific Network Information Centre
ARIN	American Registry for Internet Numbers
ASO	(ICANN) Address Supporting Organization
BITs	Bilateral Investment Treaties
BRIC	Brazil, Russian Federation, India, and China
BRICS	Brazil, Russian Federation, India, China, and Republic of South Africa
CAS	Copyright Alert System
ccNSO	(ICANN) Country Code Names Supporting Organization
ccTLD	Country Code Top-Level Domain
CDEUNDP	Capacity Development Expert in United Nation Development Programme
CETA	Comprehensive Economic and Trade Agreement
CFSP	Common Foreign and Security Policy
CIA	Central Intelligence Agency
CIP	Critical Infrastructure Protection

CIS	Centre for Internet and Society
CJEU	Court of Justice of the European Union
CNCI	Comprehensive National Cyber Security Initiative
CoE	Council of Europe
COP 21	21st Conference of the Parties to the United Nations Framework Convention on Climate Change
CSDP	Common Security and Defence Policy
CST/FVG	Center for Technology and Society of the Getulio Vargas Foundation
CTD	Center for Democracy & Technology
DARPA	(US) Defense Advanced Research Projects Agency
DEA	(UK) Digital Economy Act 2010
DMCA	(US) Digital Millennium Copyright Act
DNS	Domain Name System
DoC	(US) Department of Commerce
DPD	(EU) Data Protection Directive 1995
EC	European Commission
ECHR	European Convention on Human Rights
EDPS	European Data Protection Supervisor
EEAS	European External Action Service
ELAC	Economic Commission for Latin America and the Caribbean
ENISA	European Union Agency for Network and Information Security
ESCWA	United Nation Economic and Social Commission for Western Asia
ETSI	European Telecommunications Standards Institute
FANCV	Fundación Argentina a las Naciones Camino a la Verdad
FAO	Food and Agriculture Organization
FCC	(US) Federal Communications Commission
FTA	Free Trade Agreement
G20	Group of Twenty, a forum for the governments and central banks of nineteen nations and the European Union founded in 1999
G22	Group of Twenty-Two, a coalition of the G8 and 14 nations mainly from Asia and Latin America formed in 1997, initially intended to reform the global financial system
G77	Group of Seventy-Seven, a coalition of developing nations initially formed by seventy-seven founding members in 1964 at the UNCTAD

GAC	(ICANN) Government Advisory Committee
GAFA	Google, Apple, Facebook, and Amazon
GATS	General Agreement on Trade in Services
GATT	General Agreement on Tariffs and Trade
GCSC	Global Commission on Stability in Cyberspace
GDPR	(EU) General Data Protection Regulation
GGE	(UN) Group of Governmental Experts on Advancing responsible State behaviour in cyberspace in the context of international security
GNSO	(ICANN) Generic Names Supporting Organization
GPD	Global Partners Digital
GSM	Global System for Mobile Communication
gTLD	Generic Top-Level Domain
HADOPI	(French) Haute Autorité pour la diffusion des œuvres et la protection des droits sur Internet
IACHR	Inter-American Commission on Human Rights
IANA	(ICANN) Internet Assigned Numbers Authority
ICANN	Internet Corporation for Assigned Names and Numbers
ICC	International Chamber of Commerce
ICRC	International Committee of the Red Cross
ICT	Information and Communication Technologies
IDP	Internet Democracy Project
IEEE	Institute of Electrical and Electronic Engineers
IETF	Internet Engineering Task Force
IFLA	International Federation of Library Associations and Institutions
IFTA	Independent Film and Television Alliance
IG	Internet Governance
IGF	Internet Governance Forum
IGO	Intergovernmental Organization
ILETS	International Law Enforcement Telecommunications Seminar
ILO	International Labor Organization
IMF	International Monetary Fund
IMO	International Maritime Organization
INAI	(Mexican) Instituto Nacional de Transparencia, Acceso a la Información y Protección de Datos Personales (formerly IFAI: Instituto Federal de Acceso a la Información y Protección de Datos)
IO	International Organization
IRPC	Internet Rights and Principles Coalition

ISOC	Internet Society
ISP	Internet Service Provider
ITU	International Telecommunication Union
IUR	International User Requirements
JBF	Japan Business Federation
JISA	Japan Information Technology Services Industry Association
JPNIC	Japan Network Information Center
JPRS	Japan Registry Services
KLRCA	Kuala Lumpur Regional Centre for Arbitration
LACNIC	Latin American and Caribbean IP address Regional Registry
LEA	Law Enforcement Agency
LEMF	Law Enforcement Monitoring Facility
MIKTA	Mexico, Indonesia, Republic of Korea, Turkey, and Australia
MIT	Multilateral Investment Treaty
MoU	Memorandum of Understanding
MPAA	Motion Picture Association of America
MS	Member State
MSG	Multistakeholder Group
MYNIC	Malaysia Network Information Centre
NetMundial	NetMundial Intitiative - Global Multistakeholder Meeting on the Future of Internet Governance
NGO	Non Governmental Organization
NIS	Network and Information Security
NPF	Narrative Policy Framework
NRC	National Research Council
NSA	(US) National Security Agency
NTIA	(US) National Telecommunications and Information Administration
OECD	Organisation for Economic Cooperation and Development
OEWG	(UN) Open-Ended Working Group on Developments in the Field of ICTs in the Context of International Security
OMB	(US) Office of Management and Budget
PASA	Peoples Advancement Social Association
PICDRP	Public Interest Commitment Dispute Resolution Procedure

PTI	(ICANN) Public Technical Identifiers (IANA's successor)
PTT	Postal, Telegraph, and Telephone Agency
QUANGO	Quasi (Autonomous) Non Governmental Organization
RALO	(ICANN) Regional At-Large Organization
RFC	Request for comments: a memorandum submitted for peer review (terminology mostly used by the technical community)
RIAA	Recording Industry Association of America
RIPE NCC	Réseaux IP Européens - Network Coordination Centre (RIR for Europe, Middle-East and part of Central Asia)
RIRs	Regional Internet Registries (AFRINIC, ARIN, APNIC, LACNIC and RIPE NCC)
RPZ	Response Policy Zone
S&T	Science and Technology
SDGs	Sustainable Development Goals
STAS	(US) Office of the Science and Technology Adviser to the Secretary of State
STI	Science Technology and Innovation
TiSA	Trade in Services Agreement
TTIP	Transatlantic Trade and Investment Partnership
UN	United Nations
UNCTAD	United Nations Conference on Trade and Development
UNESCO	United Nations Educational, Scientific and Cultural Organization
UNGA	United Nations General Assembly
USAID	United States Agency for International Development
VCDR	Vienna Convention on Diplomatic Relations
WEF	World Economic Forum
WGIG	(UN) Working Group on Internet Governance
WIPO	World Intellectual Property Organization
WMO	World Meteorological Organization
WSIS	World Summit on Information Society
WSIS+10	World Summit on Information Society +10 review process
WTO	World Trade Organisation

References

Abelson, Harold, Ross Anderson, Steven Bellovin, Josh Benaloh, Matt Blaze, Whitfield Diffie, John Gilmore, et al. 2015. "Keys under Doormats: Mandating Insecurity by Requiring Government Access to All Data and Communications." Computer Science and Artificial Intelligence Laboratory Technical Report. Cambridge: MIT.

Acuto, Michele. 2013. *Global Cities, Governance and Diplomacy*. London: Routledge.

———. 2017. "On diplomacy beyond *On Diplomacy*: Time for a second edition." *New Perspectives* 25(3): 9–95.

Adamson, Fiona B. 2016. "Spaces of Global Security: Beyond Methodological Nationalism." *Journal of Global Security Studies* 1(1): 19–35.

Adler-Nissen, Rebecca. 2014. "Symbolic power in European diplomacy: the struggle between national foreign services and the EU's External Action Service", *Review of International Studies* 40(4): 657–481.

Adler-Nissen, Rebecca, and Alena Drieschova. 2019. "Track-Change Diplomacy: Technology, Affordances, and the Practice of International Negotiations." *International Studies Quarterly* 63(3): 531–45.

Aglietta, Michel. 1998. "Capitalism at the Turn of the Century: Regulation Theory and the Challenge of Social Change." *New Left Review* I (232), now in Id. 2000. *'Postface'. A Theory of Capitalist Regulation: the US Experience*. London: Verso.

Ahrne, Goran, and Nils Brunsson. 2008. *Meta-organizations*. Cheltenham: Edward Elgar.

Ahrne, Goran, and Nils Brunsson. 2012. "How Much do Meta-Organizations Affect Their Members?" In *Weltorganisationen,* edited by Martin Koch, 57–70. Wiesbaden: VS Verlag für Sozialwissenschaften.

Ahrne, Goran, Nils Brusson, and Dieter Kerwer. 2016. "The Paradox of Organizing States: A Meta-Organization Perspective on International Organizations." *Journal of Information and Organizational Sciences* 7(1): 5–24.

Aldrich, Richard. 2009. "Beyond the Vigilant State: Globalisation and Intelligence." *Review of International Studies* 35(4): 889–902.

Alicea, Tyler. 2013. "Verizon CEO Stresses Importance of Communications Technology." *The Cornell Daily Sun*, September 19, 2013. Accessed June 11, 2021. http://cornellsun.com/2013/09/19/verizon-ceo-stresses-importance-of-communications-technology.

Amoretti, Francesco, and Domenico Fracchiolla. 2022. "Modes of Internet Governance as Science Diplomacy: What Might the EU Learn from the US Cyber Security Policy?" In *Internet Diplomacy: Shaping the Global Politics of Cyberspace*, edited by M. Marzouki and A. Calderaro. Lanham: Rowman & Littlefield.

Anghie, Antonie. 2004. *Imperialism, Sovereignty, and the Making of International Law*. New York: Cambridge University Press.

Aranzales, Jhon Kelly Bonilla. 2017. "Understanding the Role of Epistemic Communities in the Development of Science Diplomacy of the United States of America and Germany towards Colombia." MSc. Diss., Universidad del Rosario, Bogotá.

Aroba, Nidhi. 2013. "Implementation and Success Analysis of Various Global Graduated Response Programs for Piracy with Special Focus on the "Six Strikes" Policy." Electronic Thesis, University of Arizona. Accessed June 11, 2021. https://repository.arizona.edu/handle/10150/297511.

Arpagian, Nicolas. 2016. "The Delegation of Censorship to the Private Sector." In *The Turn to Infrastructure in Internet Governance*, edited by F. Musiani, D. L. Cogburn, L. DeNardis and N. S. Levinson, 155–67. New York: Palgrave Macmillan.

Article 29 WP. 2010. "Opinion 2/2010 on online behavioural advertising (00909/10/EN WP 171)." Adopted on 22 June 2010 by the Working Party set up under Article 29 of European Directive 95/46/EC.

———. 2011. "Opinion 15/2011 on the definition of consent (01197/11/EN WP187)." Adopted on 13 July 2011 by the Working Party set up under Article 29 of Directive 95/46/EC.

Artigo 19. 2016 "O Dereito ao Esquecimento na América Latina." In *Libertad de Expresión en el Ámbito digital. El estado de la situación en América Latina*, edited by ADC—Asociacion por los Derechos Civiles, 46–69. Accessed June 11, 2021. https://karisma.org.co/wp-content/uploads/download-manager-files/LibEx-en-LatAm-AmbitoDigital.pdf.

Autorité de la Concurrence and Bundeskartellamt. 2016. "Competition Law and Data." Report. Accessed June 11, 2021. https://www.bundeskartellamt.de/SharedDocs/Publikation/DE/Berichte/Big%20Data%20Papier.pdf.

Bäckstrand, Karin. 2006. "Democratizing Global Environmental Governance? Stakeholder Democracy after the World Summit on Sustainable Development." *European Journal of International Relations* 12(4): 467—98.

Badie, Bertrand. 2013. "Transnationalising diplomacy and global governance." In *Diplomacy in a Globalizing World. Theories and Practices*, edited by P. Kerr and G. Wiseman, 85–102. Oxford: Oxford University Press.

Badouard, Romain, Clément Mabi, and Guillaume Sire. 2016. "Beyond 'Points of Control': Logics of Digital Governmentality." *Internet Policy Review* 5(3). https://doi.org/10.14763/2016.3.433.

Barrinha, André, and Thomas Renard. 2017. "Cyber-Diplomacy: The Making of an International Society in the Digital Age." *Global Affairs* 3(4–5): 353–64.

Barron, Anne. 2011. "'Graduated Response' *à l'Anglaise*: Online Copyright Infringement and the Digital Economy Act 2010." *Journal of Media Law* 3(2): 305–47.

Bazzicalupo, Laura. 2006. *Il governo delle vite. Biopolitica e bioeconomia*. Roma-Bari: Laterza.

———. 2015. "Radical Historicization, Genealogy of Governmentality and Political Subjectivation." *Soft Power* 2(1). Accessed June 11, 2021. http://www.softpower journal.com/web/wp-content/uploads/2016/09/SPJ03_04_LauraBazzicalupo.pdf.

Behravesh, Maysam. 2011. "The Relevance of Constructivism to Foreign Policy Analysis." *E-International Relations*. Accessed June 11, 2021. http://www.e-ir .info/2011/07/17/the-relevance-of-constructivism-to-foreign-policy-analysis/.

Bendiek, Annegret. 2014. *Tests of Partnership: Transatlantic Cooperation in Cyber Security, Internet Governance and Data Protection*. Berlin: Stiftung Wissenschaft und Politik (SWP).

Bendiek, Annegret, and Matthias C. Kettemann. 2021. "Revisiting the EU Cybersecurity Strategy: A Call for EU Cyber Diplomacy", *SWP Comment 2021(16)*, February 24, 2021. http://doi.org/10.18449/2021C16.

Benkler, Yochai, and Helen Nissenbaum. 2006. "Commons-based Peer Production and Virtue." *Journal of Political Philosophy* 14(4): 394–419.

Berke, Juergen. 1992. "Mobilfunk Lauscher überlistet." *Wirtschaftswoche*, November 11, 1992.

Berridge, G. R. 2011. *The Counter-Revolution in diplomacy and Other Essays*. Basingstoke: Palgrave Macmillan.

———. 2015. *Diplomacy. Theory and Practice* (5 ed.). Basingstoke: Palgrave Macmillan.

Berridge, G. R., and Lorna Lloyd. 2012. *The Palgrave Macmillan Dictionary of Diplomacy* (3 ed.). Basingstoke: Palgrave Macmillan.

Bevir, Mark. 2006. *Encyclopedia of Governance*. London: SAGE Publications.

Bijker, Wiebe, Roland Bal, and Ruud Hendriks. 2009. *The Paradox of Scientific Authority: The Role of Scientific Advice in Democracies*. Cambridge: MIT Press.

Bijker, Wiebe, Thomas P. Hughes, and Trevor Pinch, eds. 2012. *The Social Construction of Technological Systems: New Directions in the Sociology and History of Technology*. Cambridge: MIT Press.

Bjola, Corneliu, and Marcus Holmes. 2015. *Digital Diplomacy. Theory and Practice*. Abingdon, Oxon: Routledge.

Bjola, Corneliu, and Ruben Zaiotti, eds. 2020. *Digital Diplomacy and International Organisations: Autonomy, Legitimacy and Contestation*. London: Routledge.

Boardman, Michael. 2011. "Digital Copyright Protection and Graduated Response: A Global Perspective." *Loyola of Los Angeles International & Comparative Law Review* 33(2): 222–54.

Bobbio, Norberto. 1999. *Teoria generale della politica*, Torino: Einaudi.

Boydstone, J. A., ed. 1989. *John Dewey: The Later Works, 1925–1953, Volume 16, Essays, Typescripts, and Knowing and the Known*. Carbondale: Southern Illinois University Press.

Bozeman, Barry. 2007. *Public Values and Public Interest*. Washington DC: Georgetown University Press.

Braman, Sandra. 2004. "The Emergent Global Information Policy Regime." In *The Emergent Global Information Policy Regime*, edited by Sandra Braman, 12–37. Houndsmills: Palgrave Macmillan.

Breindl, Yana and François Briatte. 2013. "Digital Protest Skills and Online Activism Against Copyright Reform in France and the European Union." *Policy & Internet* 5(1): 27–55.

Broeders, Dennis. 2015. *The Public Core of the Internet: An International Agenda for Internet Governance*. Amsterdam: Amsterdam University Press.

Broeders, Dennis, and Bibi van den Berg. 2020a. "Governing Cyberspace: Behavior, Power, and Diplomacy." In *Governing Cyberspace: Behavior, Power, and Diplomacy*, edited by Dennis Broeders and Bibi van den Berg, 1–17. Lanham: Rowman & Littlefield.

Broeders, Dennis, and Bibi van den Berg, eds. 2020b. *Governing Cyberspace. Behavior, Power and Diplomacy*. London: Rowman & Littlefield.

Brousseau, Eric, and Meryem Marzouki. 2012. "Old Issues, New Framings, Uncertain Implications". In *Governance, Regulations and Powers on the Internet*, edited by E. Brousseau, M. Marzouki and C. Meadel, 368–97. Cambridge: Cambridge University Press.

Brousseau, Eric, Meryem Marzouki, and Cécile Méadel. 2012a. "Governance, Networks and Digital Technologies: Societal, Political and Organizational Innovations." In *Governance, Regulations and Powers on The Internet*, edited by E. Brousseau, M. Marzouki and C. Méadel, 3–36. Cambridge: Cambridge University Press.

Brousseau, Eric, Meryem Marzouki, and Cécile Méadel, eds. 2012b. *Governance, Regulations and Powers on The Internet*. Cambridge: Cambridge University Press.

Brown, Ian, and Christopher T. Marsden. 2013 *Regulating Code: Good Governance and Better Regulation in the Information Age*. Cambridge: MIT Press.

Brunner, Lisl. 2016. "The Liability of an Online Intermediary for Third Party Content: the Watchdog Becomes the Monitor: Intermediary Liability after Delfi v Estonia." *Human Rights Law Review* 16(1): 163–74.

Bunyan, Tony, Heiner Busch, Elspeth Guild, and Steve Peers. 2000. *The Impact of the Amsterdam Treaty on Justice and Home Affairs Issues*. Civil Liberties working paper series. Brussels: European Parliament Directorate-General for Research. Accessed June 11, 2021. http://www.europarl.europa.eu/RegData/etudes/etudes/join/2000/228145/IPOL-LIBE_ET(2000)228145_EN.pdf.

Burchell, Graham, Gordon Colin, and Miller Peter, eds. 1991. *The Foucault Effect: Studies in Governmentality*. Chicago: University of Chicago Press.

Butt, Danny. 2016. "FCJ–198 New International Information Order (NIIO) revisited: Global Algorithmic Governance and Neocolonialism." *Fibreculture Journal* 27: 1–37.

Bygrave, Lee A. 2013. "Transatlantic Tensions on Data Privacy." *Transworld Papers* n°19, April 8, 2013. Accessed June 11, 2021. https://www.iai.it/sites/default/files/TW_WP_19.pdf

Calderaro, Andrea. 2021. "Diplomacy and Responsibilities in the Transnational Governance of the Cyber Domain." In *The Routledge Handbook of Responsibility in World Politics*, edited by H. Hansen-Magnusson and A. Vetterlein. London: Routledge.

Calderaro, Andrea, and Anthony J. S. Craig. 2020. "Transnational Governance of Cybersecurity: Policy Challenges and Global Inequalities in Cyber Capacity Building." *Third World Quarterly* 41(6): 917–38.

Calderaro, Andrea, and Anastasia Kavada. 2013. "Challenges and Opportunities of Online Collective Action for Policy Change." *Policy & Internet* 5(1): 1–6.

Callon, Michel. 2003. "The Increasing Involvement of Concerned Groups In R&D Policies: What Lessons for Public Powers?" In *Science and Innovation: Rethinking the Rationales for Funding and Governance*, edited by A. Geuna, A. J. Salter, and W. E. Steinmueller, 30–68. Cheltenham: Edward Elgar Publishing.

Callon, Michel, and Vololona Rabeharisoa. 2003. "Research 'In the Wild' and the Shaping of Social Identities." *Technology in Society* 25(2): 193–204.

Cammaerts, Bart, and Robin Mansell. 2020. "Digital Platform Policy and Regulation: Toward a Radical Democratic Turn." *International Journal of Communication* 14: 1–19.

Campaign for Accountability. 2016. *Google's Silicon Tower. Google-Funded Speakers Dominate Key 2016 Policy Conferences.* Accessed June 11, 2021. https://assets.documentcloud.org/documents/2995349/CfA-Googles-Silicon-Tower-7-19-16-Final.pdf

———. 2017. *Google Academics Inc.* Accessed June 11, 2021. http://googletransparencyproject.org/sites/default/files/Google-Academics-Inc.pdf.

Campbell, Duncan. 1999. "The State of the Art in Communications Intelligence (COMINT) of Automated Processing for Intelligence Purposes of Intercepted Broadband Multilanguage Leased or Common Carrier Systems, and Its Applicability to COMINT Targeting and Selection, Including Speech Recognition." In *Development of Surveillance Technology and Risk of Abuse of Economic Information.* Scientific and Technological Options Assessment. Luxemburg: European Parliament. Accessed June 11, 2021. https://www.duncancampbell.org/menu/surveillance/echelon/IC2000_Report%20.pdf.

Canário, Pedro. 2016. Só História pode decidir o que deve ser esquecido, defende entidade de jornalistas. *Consultor Jurídico.* November 7, 2016. Accessed June 11, 2021. http://www.conjur.com.br/2016-nov-07/historia-quem-decide-esquecido-defende-entidade.

Cardon, Dominique. 2010. *La démocratie Internet. Promesses et limites.* Paris: Seuil/La République des Idées.

Caretti, Paolo. 2005. *I diritti fondamentali. Libertà e diritti sociali.* Torino: Giappichelli.

Carr, James. 1981. "Wiretapping in West Germany." *American Journal of Comparative Law* 29(4): 607–45.

Carr, Madeline. 2016. "Public–Private Partnerships in National Cyber-Security Strategies." *International Affairs* 92(1): 43–62.

Casilli, Antonio. 2015. "Travail, technologies et conflictualités." In *Qu'est-ce que le digital labor?*, edited by Antonio Casilli and Dominique Cardon. Créteil : Editions de l'INA.

Catania, Alfonso. 2008. *Metamorfosi del diritto. Decisione e norma nell'età globale.* Roma-Bari: Laterza.

Cavelty, Myriam Dunn. 2013. "A Resilient Europe for an Open, Safe and Secure Cyberspace." *UI Occasional Papers* no. 23. http://dx.doi.org/10.2139/ssrn.2368223.

————. 2015. "The Normalization of Cyber-International Relations." In *Strategic Trends 2015: Key Developments in Global Affairs*, edited by O. Thränert and M. Zapfe. Zurich: ETH Center for Security Studies.

Cerf, Vinton G. 2020. "On Digital Diplomacy". *Communications of the ACM* 63(10): 5.

Chakravartty Paula, and Sarikakis, Katharine. 2006. *Media Policy and Globalisation.* Edinburgh: Edinburgh University Press.

Chase, Peter, and Jacques Pelkmans. 2015. "This Time it's Different: Turbo-Charging Regulatory Cooperation in TTIP." CEPS Special Report. Accessed June 11, 2021. https://www.ceps.eu/system/files/SR110%20Regulatory%20Cooperation%20in%20TTIP.pdf.

Chenou, Jean-Marie. 2010. "Multistakeholderism or Elitism? The Creation of a Transnational Field of Internet Governance." GigaNet: Global Internet Governance Academic Network, Annual Symposium 2010. Accessed June 11, 2021. http://dx.doi.org/10.2139/ssrn.2809217.

————. 2014. "From Cyber-Libertarianism to Neoliberalism: Internet Exceptionalism, Multi-stakeholderism, and the Institutionalisation of Internet Governance in the 1990s." *Globalizations* (11)2: 205–23.

————. 2022. "In This Bright Future You Can't Forget Your Past: Debating the 'Right to Be Forgotten' in Latin America." In *Internet Diplomacy: Shaping the Global Politics of Cyberspace*, edited by M. Marzouki and A. Calderaro. Lanham: Rowman & Littlefield.

Chenou, Jean-Marie, and Radu, Roxana. 2017. "The 'Right To Be Forgotten': Negotiating Public and Private Ordering in the European Union." *Business and Society* 58(1):74–102.

Chomsky, Noam. 2016. *Who Rules the World?* New York: Metropolitan Books/Henry Holt & Co.

Chossudovsky, Michel. 2003. *The Globalization of Poverty and the New World Order.* Pincourt: Global Research.

Christakis, Theodore. 2020. *"European Digital Sovereignty": Successfully Navigating Between the "Brussels Effect" and Europe's Quest for Strategic Autonomy.* Grenoble: Multidisciplinary Institute on Artificial Intelligence/Grenoble Alpes Data Institute. Accessed June 11, 2021. http://dx.doi.org/10.2139/ssrn.3748098.

Christensen, Clayton M. 1997. *The Innovator's Dilemma: When New Technologies Cause Great Firms to Fail.* Boston: Harvard Business Review Press.

Christou, George. 2016. *Cybersecurity in the European Union Resilience and Adaptability in Governance Policy.* London: Palgrave Macmillan.

Chussudovsky, Michel. 2008. "Iran: War or Privatization: All Out War or "Economic Conquest"?" *Global Research*, July 4, 2008. Accessed June 11, 2021.

http://www.globalresearch.ca/iran-war-or-privatization-all-out-war-or-economic-conquest/9501.

Cicourel, Aaron. 1988. "Text and Context: Cognitive, Linguistic and Organizational Dimensions in International Negotiations." *Negotiation Journal* 4 (3): 257–66.

CIS. The Center for Internet and Society. n.d. Intermediary Liability. Accessed June 11, 2021. http://cyberlaw.stanford.edu/focus-areas/intermediary-liability.

CJEU. Court of Justice of the European Union. 2014. Google Spain SL, Google Inc. v. Agencia Española de Protección de Datos (AEPD), Mario Costeja González. C131/12. May 13, 2014. Accessed June 11, 2021. http://curia.europa.eu/juris/document/document_print.jsf?doclang=EN&docid=152065.

———. 2015. Judgement of the Court (Grand Chamber) of 6 October 2015. Maximillian Schrems v Data Protection Commissioner Request for a preliminary ruling from the High Court (Ireland) Judgment ECLI:EU:C:2015:650. Luxembourg: European Court of Justice. Accessed June 11, 2021. http://curia.europa.eu/juris/liste.jsf?num=C-362/14#.

Clarke, Richard A., and Robert Knake. 2011. *Cyber War: The Next Threat to National Security and What to Do About It.* New York: HarperCollins.

Claussen Victor. 2018. "Fighting Hate Speech and Fake News. The Network Enforcement Act (NetzDG) in Germany in the context of European legislation", *Media Laws—Rivista di diritto dei Media*, 2018(3): 110–36.

Clinton, Hillary. 2010. "Remarks on Internet Freedom." *US Secretary of State Remarks given at the Washington DC Newseum,* January 21, 2010. https://2009-2017.state.gov/secretary/20092013clinton/rm/2010/01/135519.htm.

Coaffee, John, and Pete Fussey. 2015. "Constructing Resilience Through Security and Surveillance: The Politics, Practices and Tensions of Security-Driven Resilience." *Security Dialogue* 46(1):86–105.

Elder, Charles D., and Roger W. Cobb. 1983. *The Political Uses of Symbols.* New York: Longman.

CoE. 1950. *Convention for the Protection of Human Rights and Fundamental Freedoms as amended by Protocols No. 11 and No. 14.* Rome: Council of Europe.

———. 1981. *Convention for the Protection of Individuals with regard to Automatic Processing of Personal Data (European Treaty Series—No. 108).* Strasbourg: Council of Europe.

———. 2010. "The protection of individuals with regard to automatic processing of personal data in the context of profiling." *Recommendation CM/Rec(2010)13 adopted by the Committee of Ministers of the Council of Europe on 23 November 2010 and explanatory memorandum.* Strasbourg: Council of Europe.

———. 2013. "Declaration of the Committee of Ministers on Risks to Fundamental Rights stemming from Digital Tracking and other Surveillance Technologies." *Adopted by the Committee of Ministers on June 11, 2013.* Strasbourg: Council of Europe.

Cohen, Michael D., James G. March, and Johan P. Olsen. 1972. "A Garbage Can Model of Organisation Choice." *Administration Science Quarterly* 17(1): 1–25.

Colglazier, E. William, and Elizabeth Lyons. 2014. "The United States Looks to the Global Science, Technology, and Innovation Horizon." *Science & Diplomacy,* July 8,

2014. Accessed June 11, 2021. https://www.sciencediplomacy.org/perspective/2014/united-states-looks-global-science-technology-and-innovation-horizon.

Collin, Pierre, and Nicolas Colin. 2013. Rapport de *Mission d'expertise sur la fiscalité de l'économie numérique*. Paris: Ministère de l'Economie et des finances. Accessed June 11, 2021. https://www.economie.gouv.fr/files/rapport-fiscalite-du-numerique_2013.pdf.

Commission on Global Governance. 1995. *Our Global Neighborhood. The Report of the Commission on Global Governance*. Accessed June 11, 2021. http://www.gdrc.org/u-gov/global-neighbourhood/.

Comor, Edward. 1999. "Governance and the Nation State in a Knowledge-Based Political Economy." In *Approaches to Global Governance Theory*, edited by M. Hewson and T. J. Sinclair, 117–34. New York: State University of New York Press.

Comor, Edward, and Hamilton Bean. 2012. "America's 'Engagement' Delusion: Critiquing a Public Diplomacy Consensus." *International Communication Gazette* 74(3): 203–20.

Cooper, Andrew F., Jorge Heine, and Ramesh Chandra Thakur, eds. 2013a. *The Oxford Handbook of Modern Diplomacy*. Oxford: Oxford University Press.

Cooper, Andrew F., Jorge Heine, and Ramesh Thakur. 2013b. "Introduction: The Challenges of 21st-Century Diplomacy." In *The Oxford Handbook of Modern Diplomacy*, edited by A. F. Cooper, J. Heine, and R. Thakur. Oxford: Oxford University Press.

Cooper, Andrew F., Brian Hocking, and William Maley. 2008. "Introduction: Diplomacy and Global Governance: Locating Patterns of (Dis)Connection." In *Global Governance and Diplomacy. Worlds Apart?*, edited by A. F. Cooper, B. Hocking and W. Maley, 1–12. Basingstoke: Palgrave Macmillan.

Copeland, Daryl. 2016. "Science Diplomacy." In *Sage Handbook of Diplomacy*, edited by C. M. Constantinou, P. Kerr, and P. Sharp, 628–41. London: Sage.

Cornut, Jérémie. 2018. "Diplomacy, Agency, and the Logic of Improvisation and Virtuosity in Practice." *European Journal of International Relations* 24(3): 712–36.

Correio Braziliense. 2016. "STF julgará ação que pode regulamentar direito ao esquecimento". *Correio Braziliense*, July 24, 2016. Accessed June 11, 2021. https://www.correiobraziliense.com.br/app/noticia/brasil/2016/07/24/interna-brasil,541424/stf-julgara-acao-que-pode-regulamentar-direito-ao-esquecimento.shtml.

Corte Constitucional. 2015. Acción de tutela instaurada por Gloria contra la Casa Editorial El Tiempo. Sentencia T–277/15. May 12, 2015. Accessed June 11, 2021. http://www.corteconstitucional.gov.co/relatoria/2015/t-277-15.htm.

Corte Suprema de Justicia de la Nación. 2014. Rodriguez, Maria Belén c/ Google Inc. S/ daños y perjuicios. R.522. XLIX. October, 28, 2014. Accessed June 11, 2021. http://www.telam.com.ar/advf/documentos/2014/10/544fd356a1da8.pdf.

Couldry, Nick, and Ulises A. Mejias. 2019. *The Costs of Connection: How Data Is Colonizing Human Life and Appropriating It for Capitalism*. Stanford: Stanford University Press.

Council of the European Union. 1995. *Council Resolution of 17 January 1995 on the Lawful Interception of Telecommunications*. Brussels: Council of the

European Union. Accessed June 11, 2021. https://eur-lex.europa.eu/legal-content/EN/TXT/?uri=CELEX%3A31996G1104.

Council of the European Union. 2015. *Council Conclusions on Cyber Diplomacy*. Brussels: Council of the European Union. Accessed June 11, 2021. http://data.consilium.europa.eu/doc/document/ST-6122-2015-INIT/en/pdf.

Cull, Nicholas J. 2013. "The Long Road to Public Diplomacy 2.0: The Internet in US Public Diplomacy." *International Studies Review* 15: 123–39.

Cummings, Richard H. 2009. *Cold War Radio: The Dangerous History of American Broadcasting in Europe, 1950–1989*. Jefferson: McFarland & Company.

Custers, Bart, Simone van der Hof, and Bart Schermer. 2014. "Privacy Expectations of Social Media Users: The Role of Informed Consent in Privacy Policies." *Policy & Internet* 6(3): 268–95.

Daly, Angela. 2016. *Private Power, Online Information Flows and EU Law*. London: Hart Publishing.

Danaher, Brett, Michael D. Smith, Rahul Telang, and Siwen Chen. 2014. "The Effect of Graduated Response Anti-Piracy Laws on Music Sales: Evidence from an Event Study in France". *Journal of Industrial Economics* LXII(3): 541–53.

Danneels, Erwin. 2004. "Disruptive Technology Reconsidered: A Critique and Research Agenda." *The Journal of Product and Innovation Management* 21(4): 246–58.

Dardot, Pierre, and Christian Laval. 2015. *Common: On Revolution in the 21st Century*. London: Bloomsbury.

———. 2017. *The New Way of the World: On Neo-Liberal Society*. London:Verso.

David, Paul A., and W. Edward Steinmueller. 2013. "Scientific Data Commons Underutilisation: Causes, Consequences and Remedial Strategies." Paper Prepared for the International Society for New Industrial Economics (ISNIE) 2013, Panel 4.E in honour of Elinor Ostrom: Governing Networked Knowledge Commons. Accessed June 11, 2021. https://dlc.dlib.indiana.edu/dlc/bitstream/handle/10535/9066/david_steinmueller.pdf?sequence=1.

David, Paul A., Matthijs den Besten, and Ralph Schroeder. 2010. "Will E-Science be Open Science?" In *World Wide Research: Reshaping the Sciences and Humanities-wide Research*, edited by W. H. Dutton and P. W. Jeffreys, 299–316. Cambridge: MIT Press.

Davidson, William, and Joseph Montville. 1981. "Foreign policy according to Freud." *Foreign Policy* 45: 145–57.

De Carolis, Massimo. 2017. *Il rovescio della libertà: tramonto del neoliberalismo e disagio della civiltà*. Macerata: Quodlibet.

De Filippi, Primavera, and Danièle Bourcier. 2013. "La double face de l'Open data." *Les Petites Affiches* 203:6–10. Accessed June 11, 2021. https://hal.archives-ouvertes.fr/hal-01026107/document.

De Minico, Giovanna. 2012. *Internet. Regola o anarchia*. Napoli: Jovene.

———. 2016. *Costituzione. Emergenza e terrorismo*. Napoli: Jovene.

De Tullio, Maria Francesca. 2016. "La *privacy* e i *big data* verso una dimensione costituzionale collettiva." *Politica del diritto* 47(4): 637–96.

De Tullio, Maria Francesca, and Giuseppe Micciarelli. 2022. "Trade Agreements and Internet Governance: Data Flow and Politics in the TiSA's Governmental Rationality." In *Internet Diplomacy: Shaping the Global Politics of Cyberspace*, edited by M. Marzouki and A. Calderaro. Lanham: Rowman & Littlefield.

Deflem, Mathieu. 2002. *Policing World Society: Historical Foundations of International Police Cooperation*. Oxford: Oxford University Press.

———. 2006. "Global Rule of Law or Global Rule of Law Enforcement? International Police Cooperation and Counterterrorism." *Annals of the American Academy of Political and Social Science* 603(1): 240–51.

Deibert, Ronald, and Rafal Rohozinski. 2010. "Risking Security: Policies and Paradoxes of Cyberspace Security." *International Political Sociology* 4(1): 15–32.

Deleuze, Gilles. 1992 "Postscript on the Societies of Control." *October* 59(Winter): 3–7.

Dempsey, James. 1997. "Communications Privacy in the Digital Age: Revising the Federal Wiretap Laws to Enhance Privacy." *Albany Law Journal of Science & Technology* 8(1): 65–120.

DeNardis, Laura. 2014. *The Global War for Internet Governance*. New Haven: Yale University Press.

———. 2020. *The Internet in Everything: Freedom and Security in a World with No Off Switch,* New Haven: Yale University Press.

DeNardis, Laura, ed. 2018. *Global Internet Governance.* London: Routledge.

DeNardis, Laura, and Andrea M. Hackl. 2015. "Internet Governance by Social Media Platforms." *Telecommunications Policy* 39(9): 761–70.

DeNardis, Laura, and Francesca Musiani. 2016. "Governance by Infrastructure." In *The Turn to Infrastructure in Internet Governance*, edited by F. Musiani, D. L. Cogburn, L. DeNardis, and N. S. Levinson, 3–21. New York: Palgrave Macmillan.

Der Derian, James. 1987. "Mediating Estrangement: A Theory for Diplomacy." *Review of International Studies* 13(2): 91–110.

———. 1999. "A Virtual Theory of Global Politics, Mimetic War, and the Spectral State." *Proceedings of the American Society of International Law* 93(January): 163–76.

———. 2017. "'Every Dog Got His Day': On Diplomacy after Thirty Years." *New Perspectives* 25(3): 73–78.

Dezalay, Yves, and Bryant G. Garth. 2002. *The Internationalization of Palace Wars: Lawyers, Economists, and the Contest to Transform Latin American States*. Chicago: University of Chicago Press.

Di Camillo, Federica, and Valérie Miranda. 2011. "Ambiguous Definitions in the Cyber Domain: Costs, Risks, and the Way Forward." *IAI Working Papers* 1126, September 14, 2011. https://doi.org/978-88-98042-32-6.

Diamond, Jeremy. 2016. "Trump, the Computer and Email Skeptic-in-Chief." CNN, December 30, 2016. Accessed June 11, 2021. https://edition.cnn.com/2016/12/29/politics/donald-trump-computers-internet-email/index.html.

Diffie, Whitfield, and Susan Landau. 2009. "Communications Surveillance: Privacy and Security at Risk." *Communications of the ACM* 52(11): 42–47.

Dingwerth, Klaus. 2014. "Global Democracy and the Democratic Minimum: Why a Procedural Account Alone is Insufficient." *European Journal of International Relations* 20(4): 1124–47.

Diplo Foundation. 2016. *Towards a Secure Cyberspace Via regional Co-Operation. Overview of the main diplomatic instruments.* Geneva: Diplo Foundation. Accessed June 11, 2021. https://www.diplomacy.edu/sites/default/files/Diplo%20 -%20Towards%20a%20secure%20cyberspace%20-%20GGE.pdf.

Dobusch, Leonhard. 2009. "Copyright Related Social Movements: Pirate Parties and the European Parliamentary Elections." *Governance Across Borders,* June 8, 2009. Accessed June 11, 2021. https://governancexborders.com/2009/06/08/copy right-related-social-movements-pirate-parties-and-the-european-parliamentary -elections/.

Dobusch, Leonhard, and Sigrid Quack. 2008. "Epistemic Communities and Social Movements. Transnational Dynamics in the Case of Creative Commons." *MPIfG Discussion Paper* 08/8. Accessed June 11, 2021. http://www.mpifg.de/pu/mpifg_ dp/dp08-8.pdf.

Domingues, Amanda, and P. H. R. Neto,. 2017. "Science Diplomacy as a Tool of International Politics: the Power of 'Soft Power.'" *Brazilian Journal of International Relations* 6(3): 607–29.

Doria, Avri 2014. "Use [and Abuse] of Multistakeholderism in the Internet." In *The Evolution of Global Internet Governance. Principles and Policies in the Making,* edited by R. Radu, J.-M. Chenou, and R. H. Weber, 115–38. Berlin: Springer.

Dragu, Tiberiu. 2011. "Is There a Trade-off Between Security and Liberty? Executive Bias, Privacy Protections, and Terrorism Prevention." *American Political Science Review* 105(1): 64–78.

Drake, William J. 2004. "Reframing Internet Governance Discourse: Fifteen Baseline Propositions." In *Internet Governance: A Grand Collaboration,* edited by D. MacLean, 122–661. New York: United Nations Information and Communication Technology Taskforce.

Drexl, Josef. 2016. "Designing Competitive Markets for Industrial Data—Between Propertisation and Access." Max Planck Institute for Innovation & Competition Research Paper No. 16–13. http://dx.doi.org/10.2139/ssrn.2862975.

Drezner, Daniel W. 2004. "Global Governance of the Internet: Bringing the State Back In." *Political Science Quarterly* 119(3): 477–98.

Drye, Kelly. 2017. "Congress Repeals FCC 2016 Privacy Order via Congressional Review Act.", April 4, 2017. Accessed June 11, 2021. https://www.lexology.com/ library/detail.aspx?g=53ed4c29-821c-45c8-8c14-fe2601d2ba90.

Dufour, Paul. 2016. "Carnegie Group at Twenty-Five: Diplomacy and Science at a High Level." *Science & Diplomacy*, February 29, 2016. Accessed June 11, 2021. https://www.sciencediplomacy.org/article/2016/carnegie-group-twenty-five.

Dutton, William. 2017. "Twitter Foreign Policy and the Rise of Digital Diplomacy." Oxford Internet Institute blog, January 5, 2017. Accessed June 11, 2021. https:// www.oii.ox.ac.uk/blog/twitter-foreign-policy-and-the-rise-of-digital-diplomacy/.

Ebrahimian, Laleh. 2003. "Socio-economic Development in Iran Through Information and Communications Technology." *Middle East Journal* 57(1): 93–111.

EC. 2009. "Agreement on EU Telecoms Reform paves way for stronger consumer rights, an open internet, a single European telecoms market and high-speed internet connections for all citizens." *EC MEMO/09/491*, November 5, 2009. Brussels: European Commission. Accessed June 11, 2021. http://europa.eu/rapid/press -release_MEMO-09-491_en.htm.

———. 2014. COM/2014/072 final. *Internet Policy and Governance: Europe's Role in Shaping the Future of Internet Governance.* Brussels: European Commission.

———. 2016. Decision 2016/1250. *EU-US Privacy Shield Agreement.* Brussels. European Commission.

———. 2017a. COM(2017)10 final. *Proposal for a Regulation of The European Parliament and of the Council Concerning the Respect for Private Life and the Protection of Personal Data in Electronic Communications and Repealing Directive 2002/58/EC.* Brussels: European Commission.

———. 2017b. *Cybersecurity in the European Digital Single Market.* High Level Group of Scientific Advisors, Scientific Opinion No. 2/2017, March 24, 2017. Brussels: European Commission.

———. 2020. COM(2020)825 final. *Proposal for a Regulation of the European Parliament and of the Council on a Single Market for Digital Services (Digital Services Act).* Brussels: European Commission.

Edelma, Murray. 1988. *Constructing the Political Spectacle.* Chicago: Chicago University Press.

EDPS. 2015. Opinion 7/2015. *Meeting the Challenges of Big Data.* Brussels: European Data Protection Supervisor.

Eggerton, John 2016. "Judge Rejects AG Attempt to Block IANA transition", *Broadcasting + Cable,* October 1, 2016. Accessed June 11, 2021. https://www.nexttv .com/news/judge-rejects-ag-attempt-block-iana-transition-160061.

Einstein, Albert. 1934/2009. "Address at Columbia University, New York, 15 January." In *Albert Einstein: Einstein's Essays in Science*, edited by A. Harris, 112–13. New York: Dover Publications.

Elkin-Koren, Niva, and Eldar Haber. 2016. "Governance by Proxy: Cyber Challenges to Civil Liberties." *Brooklyn Law Review* 82(1): 105–62.

ENISA. 2014. "An evaluation Framework for National Cyber Security Strategies." *ENISA,* November 27, 2014. Heraklion: European Union Agency for Cybersecurity.

EP. 2009. 2007/0247(COD). *Electronic Communications: Common Regulatory Framework for Networks and Services, Access, Interconnection and Authorisation. 'Telecoms Package'.* Brussels/Strasbourg: European Parliament.

EPIC. 2002. *Privacy and Human Rights: An International Survey of Privacy Laws and Developments.* Washington DC: Electronic Privacy Information Center.

Epstein, Dmitry, Christian Katzenback, and Francesca Musiani. 2016. "Doing Internet Governance: Practices, Controversies, Infrastructures, and Institutions." *Internet Policy Review* 5(3): 1–14.

Epstein, Dmitry, Merrill C. Roth, and Eric P.S. Baumer. 2014. "It's the Definition, Stupid! Framing of Online Privacy in the Internet Governance Forum Debates." *Journal of Information Policy* 4: 144–72.

EU. 1995. *Directive 95/46/EC of the European Parliament and the Council of 24 October 1995 on the Protection of Individuals with Regard to the Processing of Personal Data and on the Free Movement of Such Data*. Brussels: European Union.

———. 2000. *Directive 2000/31/EC of the European Parliament and of the Council of 8 June 2008 on Certain Legal Aspects of Information Society Services, in Particular Electronic Commerce, in the Internal Market*. Brussels: European Union.

———. 2002. *Directive 2002/58/EC of the European Parliament and of the Council of 12 July 2002 Concerning the Processing of Personal Data and the Protection of Privacy in the Electronic Communications*. Brussels: European Union.

———. 2012. 2012/C 326/02. *Charter of Fundamental Rights of the European Union*. Brussels: European Union.

———. 2016. *Regulation (EU) 2016/679 of the European Parliament and of the Council of 27 April 2016 on the Protection of Natural Persons with regard to the Processing of Personal Data and on the Free Movement of Such Data, and Repealing Directive 95/46/EC ('GDPR')*. Brussels: European Union.

Euractiv. 2009. "EU Telecoms Reform Package Agreed". *Euractiv*, November 5, 2009. Accessed June 11, 2021. http://www.euractiv.com/section/digital/news/eu-telecoms-reform-package-agreed/.

European Union Agency for Fundamental Rights and Council of Europe. 2014. *Handbook on European Data Protection Law*. Luxembourg: Publications Office of the European Union.

Fähnrich, Birte. 2017. "Science diplomacy: Investigating the Perspective of Scholars on Politics–Science Collaboration in International Affairs." *Public Understanding of Science* 26(6): 688–703.

Farrell, Henry, and Abraham L. Newman. 2019. "Weaponized Interdependence: How Global Economic Networks Shape State Coercion." *International Security* 44(1): 42–79.

FBI—Federal Bureau of Investigation. 1994. "Law Enforcement Requirements for the Surveillance of Electronic Communications." In *Cryptography and Privacy Sourcebook, 1995: Documents on Encryption Policy, Wiretapping, and Information Warfare*, edited by D. Banisar, 51–79. Washington DC: DIANE Publishing.

FCC. 2015. "In the Matter of Protecting and Promoting the Open Internet," *Report and Order on Remand, Declaratory Ruling, and Order*. GN Docket No. 14–28. Washington DC: Federal Communication Commission.

———. 2016. "FCC Adopts Privacy Rules to Give Broadband Consumers Increased Choice, Transparency and Security of Their Personal Data, Report and Order FCC 16–106." *FCC News,* 27 October 2016. Washington DC: Federal Communications Commission.

Ferrarese, Maria Rosaria. 2000. *Le istituzioni della globalizzazione. Diritto e diritti nella società transnazionale*. Bologna: Il Mulino.

———. 2014. "Sulla governance paragiudiziaria. Arbitrati e investimenti esteri." *Politica del diritto* 45(3): 375–402.

———. 2017. "Governance, arbitrati ed economia politica". In *Al di là del nesso autorità/libertà: tra legge e amministrazione*, edited by S. Perongini. Torino: Giappichelli.

Ferrari, Verónica, and Schnidrig, Daniela. 2015. *Responsabilidad de Intermediarios y derecho al Olvido. Aportes para la discusión legislativa en Argentina.* Centro de Estudios en Libertad de Expresión y Acceso a la Información, Universidad de Palermo. Accessed June 11, 2021. http://www.palermo.edu/cele/pdf/Policy_Paper_Derecho_al_Olvido.pdf

Ferti, Andriani. (2010). "The Internet Freedom Provision of the EU Telecoms Package." *Computer Law Review International* 11(1):1–8.

Fidler, David P. 2011. "Navigating the Global Health Terrain: Mapping Global Health Diplomacy." *Asian Journal of WTO & International Health Law and Policy* 6(1): 1–43.

Fioravanti, Maurizio. 2009. *Costituzionalismo. Percorsi della storia e tendenze attuali* Roma-Bari: Laterza.

Fischer, Frank. 2003. *Reframing Public Policy: Discursive Politics and Deliberative Practices.* Oxford: Oxford University Press.

Fisher, Mark. 2009. *Capitalist Realism: Is There No Alternative.* Winchester (UK)-Washington (USA): Zero Books.

Folha de S. Paulo. 2015. "STF mantém decisão que permite busca no Google por filme erótico de Xuxa". *Folha de S. Paulo*, September 15, 2015. Accessed June 11, 2021. http://www1.folha.uol.com.br/ilustrada/2015/09/1682024-stf-mantem -decisao-que-permite-busca-no-google-por-filme-erotico-de-xuxa.shtml.

Foucault, Michel. 2000. "Omnes et Singulatim: Towards a Criticism of 'Political Reason'" [1979]. In *Power: Essential Works of Michel Foucault 1954-1984*, vol. 3, edited by J. D. Faubion. New York: The New Press.

———. 2004. *Sécurité, Territoire, Population. Cours prononcé au Collège de France 1977–1978.* Paris: Gallimard/Seuil.

Frank, Robert. 2012. "Internationalisation du sport et diplomatie sportive." In *Pour l'histoire des relations internationales*, edited by R. Frank, 387–405. Paris: Presses Universitaires de France.

Franklin, Marianne I. 2013a. *Digital Dilemmas: Power, Resistance, and the Internet.* Oxford: Oxford University Press.

———. 2013b. "Is Another Internet Possible? The Global Battle for Control of Cyberspace." ECPR General Conference 2013, September. Accessed June 11, 2021. https://ecpr.eu/Events/Event/PaperDetails/5168.

Fraser, Nancy. 2010. *Scales of Justice: Reimagining Political Space in a Globalizing World.* New York: Columbia University Press.

Freeman, Charles W. J. 1993. *Diplomat's Dictionary.* Washington DC: National Defense University Press.

French, Barry. 2010. "Statement to the Public Hearing on New Information Technologies and Human Rights" (transcript), June 2, 2010. Accessed June 11, 2021. http://web.archive.org/web/20100823052744/http://www.nokiasiemensnetworks.com/news-events/press-room/statement-to-the-public-hearing-on-new-information -technologies-and-human-rights.

G20. 2016. *G20 Digital Economy Development and Cooperation Initiative.* Cancun. Accessed June 11, 2021. https://www.mofa.go.jp/files/000185874.pdf.

Galli, Francesca. 2016. "The Interception of Communication in France and Italy—What Relevance for the Development of English Law?" *The International Journal of Human Rights* 20(5): 666–83.

Garnham, Nicholas. 2006. "Contribution to a Political Economy of Mass-Communication." In *Media and Cultural Studies: Keyworks*, edited by M. G. Durham and D. M. Kellner. Oxford: Blackwell Publishing.

GCIG. 2016. *One Internet*. Ottawa: Global Commission on Internet Governance. Accessed June 11, 2021. https://www.cigionline.org/publications/one-internet/.

GCSC. 2017. *Launch of Global Commission on the Stability of Cyberspace*. The Hague: Global Commission on the Stability of Cyberspace. Accessed June 11, 2021. https://cyberstability.org/news/launch-of-global-commission-on-the-stability-of-cyberspace/.

———. 2019. *Advancing Cyberstability: Final Report*. The Hague: Global Commission on the Stability of Cyberspace. Accessed June 11, 2021. https://cyberstability.org/wp-content/uploads/2020/02/GCSC-Advancing-Cyberstability.pdf.

Gerspacher, Nadia. 2005. "The Roles of International Police Cooperation Organizations." *European Journal of Crime, Criminal Law and Criminal Justice* 13(3): 413–34.

Giblin, Rebecca. 2014. "Evaluating Graduated Response." *Columbia Journal of Law & the Arts* 37(2): 147–210. http://dx.doi.org/10.2139/ssrn.2322516.

Giddens, Anthony. 1984. *The Constitution of Society. Outline of the Theory of Structuration*. Cambridge: Polity.

Gieryn, Thomas F. 1983. "Boundary-Work and the Demarcation of Science from Non-Science: Strains and Interests in Professional Ideologies of Scientists." *American Sociological Review* 48(6): 781–95.

Gilardi, F. 2012. "Transnational Diffusion: Norms, Ideas, and Policies." In *Handbook of International Relations*, edited by W. Carlsnaes, T. Risse-Kappen, and B. A. Simmons (Second edition, 453–77). Los Angeles: SAGE.

Gill, Stephen. 1995. "Globalisation, Market Civilisation, and Disciplinary Neoliberalism." *Millennium: Journal of International Studies* 24(3): 399–423.

Gill, Stephen, and David Law. 1988. *The Global Political Economy. Perspectives, Problems, and Policies*. Baltimore: John Hopkins University Press.

Gleckman, Harris. 2012. *Readers' Guide: Global Redesign Initiative*. Boston: University of Massachusetts Boston's Center for Governance and Sustainability. Accessed June 11, 2021. https://www.umb.edu/editor_uploads/gri_uploads/An_Overview_of_WEFs_Perspective.pdf.

———. 2016. "Multi-Stakeholder Governance: A Corporate Push For a New Form of Global Governance." In *State of Power 2016: Democracy, Sovereignty and Resistance*, edited by N. Buxton and D. Eade, 90–106. Amsterdam: Transnational Institute.

Goffman, Erving. 1983. "The Interaction Order: American Sociological Association, 1982 Presidential Address." *American Sociological Review* 48(1): 1–17.

Goldsmith, Jack, and Tim Wu. 2006. *Who Controls the Internet? Illusions of a Borderless World*. Oxford: Oxford University Press.

Grassegger, Hannes, and Mikael Krogeru. 2017. "The Data That Turned the World Upside Down." *Motherboard,* January 28, 2017. Accessed June 11, 2021. https://motherboard.vice.com/en_us/article/mg9vvn/how-ourlikes-helped-trump-win.

Greenleaf, Graham. 2016. "The TPP & Other Free Trade Agreements: Faustian Bargains for Privacy?" UNSW Law, Research Paper No. 08. http://dx.doi.org/10.2139/ssrn.2732386.

Grossi, Pierfrancesco. 1972. *Introduzione ad uno studio sui diritti inviolabili nella Costituzione italiana.* Padova: Cedam.

Haber, Eldar. 2011. "The French Revolution 2.0: Copyright and the Three Strikes Policy." *Harvard Journal of Sports & Entertainment Law* 2(2): 297–339.

HADOPI. 2017. Bulletin d'information trimestriel n°2—Février 2017—Chiffres clés de la réponse graduée. Paris: HADOPI.

Hajer, Maarten A. 1993. "Discourse Coalitions and the Institutionalisation of Practise" In *The Argumentative Turn in Policy Analysis and Planning*, edited by F. Fischer and J. Forester, 43–76. Durham/London: Duke University Press.

———. 1995. *The Politics of Environmental Discourse.* Oxford: Oxford University Press.

Hamilton, Keith, and Richard Langhorne. 1995. *The Practice of Diplomacy. Its Evolution, Theory, and Administration.* London: Routledge.

Hammerslev, Ole. 2012. "The European Union and the United States in Eastern Europe: Two Ways of Exporting Law, Expertise and State Power." In *Lawyers and the Rule of Law in an Era of Globalization*, edited by Y. Dezalay and B. Garth, 134–55. London: Routledge.

Harcourt, Alison, George Christou, and Seamus Simpson. 2020. *Global Standard Setting in Internet Governance.* Oxford: Oxford University Press.

Hayden, Craig. 2011. "The Lessons of Hyphenated Diplomacy", *USC Center on Public Diplomacy*, May 3, 2011. https://uscpublicdiplomacy.org/blog/lessons-hyphenated-diplomacy.

Heisenberg, Dorothee, and Marie-Helene Fandel. 2004. "Projecting EU Regimes Abroad: The EU Data Protection Directive as Global Standard." In *The Emergent Global Information Policy Regime*, edited by Sandra Braman, 109–29. Houndsmills: Palgrave Macmillan.

Held, David. 1995. *Democracy and the Global Order.* London: Polity Press.

Hill, Christopher, 1987. "On Diplomacy: A Genealogy of Western Estrangement, by James Der Derian: Review." *International Affairs* 64(1):104.

Hill, Richard. 2016. "Internet Governance, Multi-Stakeholder Models, and the IANA Transition: Shining Example or Dark Side?" *Journal of Cyber Policy* 1(2): 176–97.

Hockenos, Paul. 2017. "Germany is a Coal-Burning, Gas-Guzzling Climate Change Hypocrite." *Foreign Policy,* November 13, 2017. Accessed June 11, 2021. http://foreignpolicy.com/2017/11/13/germany-is-a-coal-burning-gas-guzzling-climate-change-hypocrite/.

Hocking, Brian, and Jan Melissen. 2015. "*Diplomacy in the Digital Age.*" The Hague: Clingendael Institute. https://www.clingendael.org/sites/default/files/2016-02/Digital_Diplomacy_in_the_Digital%20Age_Clingendael_July2015.pdf.

Hocking, Brian, Jan Melissen, Shaun Riordan, and Paul Sharp. 2012. *Futures for Diplomacy. Integrative Diplomacy in the 21st Century.* The Hague: Clingendael Institute. https://www.clingendael.org/sites/default/files/2016-02/20121030_research_melissen.pdf

Hofmann, Jeanette. 2007. "Internet Governance: A Regulative Idea in Flux". In *Internet Governance: An Introduction,* edited by R. K. J. Bandamutha, 74–108. Hyderabad: ICFAI University Press.

———. 2016. "Multi-stakeholderism in Internet Governance: Putting a Fiction into Practice." *Journal of Cyber Policy* 1(1): 29–49.

———. 2020. "The Multistakeholder Concept as Narrative: A Discourse Analytical Approach." In *Researching Internet Governance: Methods, Frameworks, Futures,* edited by L. DeNardis, D. L. Cogburn, N. S. Levinson and F. Musiani, 253–68. Cambridge, MA: MIT Press.

Hofmann, Jeanette, Christian Katzenback, and Kirsten Gollatz. 2014. *Between Coordination and Regulation: Conceptualizing Governance in Internet Governance,* HIIG Discussion Paper Series n°2014–4. https://dx.doi.org/10.2139/ssrn.2484463.

———. 2017. "Between Coordination and Regulation: Finding the Governance in Internet Governance." *New Media & Society* 19(9):1406–1423.

Holland, Adam, Chris Bavitz, Jeff Hermes, Andy Sellars, Ryan Budish, Michael Lambert, and Nick Decoster. 2015. "Intermediary Liability in the United States." Berkman Center for Internet & Society at Harvard University. Accessed June 11, 2021. https://clinic.cyber.harvard.edu/publication/intermediary-liability-in-the-united-states/.

Holmes, Marcus. 2013. "The Force of Face-to-Face Diplomacy: Mirror Neurons and the Problem of Intentions." *International Organization* 67(4): 829–61.

Holt, Rush. 2017. "AAAS CEO Urges Secretary of State to Fill Post of Science and Technology Adviser.", July 13, 2017. Washington DC: American Association for the Advancement of Science.

Höne, Katharina E. 2020. *The Future of (Multilateral) Diplomacy? Changes in Response to COVID-19 and Beyond.* Msida: DiploFoundation. Accessed June 11, 2021. https://www.diplomacy.edu/resources/books/future-multilateral-diplomacy-changes-response-covid-19-and-beyond.

———. 2022. "Diplomacy and Internet Governance: a Conceptual Re-assessment." In *Internet Diplomacy: Shaping the Global Politics of Cyberspace*, edited by M. Marzouki and A. Calderaro. Lanham: Rowman & Littlefield.

Höne, Katharina E., et al. 2019. *Mapping the Challenges and Opportunities of Artificial Intelligence for the Conduct of Diplomacy.* Msida, Malta. Accessed June 11, 2021. https://www.diplomacy.edu/sites/default/files/AI-diplo-report.pdf.

Hormats, Robert D. 2012. "Science Diplomacy and Twenty-First Century Statecraft." *Science & Diplomacy*, March 9, 2012. https://www.sciencediplomacy.org/perspective/2012/science-diplomacy-and-twenty-first-century-statecraft.

Horten, Monica. 2013. *A Copyright Masquerade: How Corporate Lobbying Threatens Online Freedom.* London, New York: Zed Books.

Hosein, Gus. 2013. "Revisiting Policy Laundering and Modern International Policy Dynamics." In *Research Handbook on Governance of the Internet*, edited by I. Brown, 260–76. Cheltenham: Edward Elgar Publishing.

Hotez, Peter J. 2014. "The NTDs and Vaccine Diplomacy in Latin America: Opportunities for United States Foreign Policy." *PLoS: Neglected Tropical Diseases* 8(9). https://doi.org/10.1371/journal.pntd.0002922.

Huang, Q. Elyse. 2020. "Facebook not Statebook: Defining SNS Diplomacy with Four Modes of Online Diplomatic Participation." *International Journal of Communication*, 14: 3885–3902.

Hurwitz, Justin G. 2007. "Whois WSIS; Whois IGF: The New Consensus-Based Internet Governance." Accessed June 11, 2021. http://ssrn.com/abstract=954209.

IACHR. Inter-American Commission on Human Rights. 2013. "Freedom of Expression and the Internet." *Report by the Office of the Special Rapporteur for Freedom of Expression.* OEA/Ser.L/V/II.CIDH/RELE/INF. 11/13. December 31, 2013. Accessed June 11, 2021. http://www.oas.org/en/iachr/expression/docs/reports/2014_04_08_internet_eng%20_web.pdf.

IANA. 2015. *The IANA Functions: An Introduction to the Internet Assigned Numbers Authority (IANA) Functions.* https://www.iana.org/about/informational-booklet.pdf.

———. 2016. *Stewardship of IANA Functions Transitions to Global Internet Community as Contract with U.S. Government Ends.* https://www.icann.org/en/announcements/details/stewardship-of-iana-functions-transitions-to-global-internet-community-as-contract-with-us-government-ends-1-10-2016-en.

ICANN. 2016. "Stewardship of IANA Functions Transitions to Global Internet Community as Contract with U.S. Government Ends." ICANN Announcements, October 1, 2016. Accessed June 11, 2021. https://www.icann.org/news/announcement-2016-10-01-en.

IFAI. 2015. En un hecho sin precedente, el IFAI inició un procediiento de imposición de sanciones en contra de Google México. January 27, 2015. Accessed June 11, 2021. http://inicio.ifai.org.mx/Comunicados/Comunicado%20IFAI-009-15.pdf.

Irion, Kristina, Svetlana Yakovleva, and Marija Bartl. 2016. "Trade and Privacy: Complicated Bedfellows? How to Achieve Data Protection-Proof Free Trade Agreements." *Independent study commissioned by BEUC et al., published 13 July 2016, Amsterdam, Institute for Information Law (IViR).* http://dx.doi.org/10.2139/ssrn.2877166.

ITU. 2005. WSIS-05/TUNIS/DOC/6(Rev. 1)-E. World Summit on the Information Society: *Tunis Agenda for the Information Society.* Geneva: International Telecommunications Union.

———. 2013. "Multistakeholderism in Internet Governance: The Importance of the Tunis Agenda." ITU News 2013, no.5 (June-July): 32-33. http://handle.itu.int/11.1004/020.3000/ITU011-2013-05-en.

Jackson, John H. 2006. *Sovereignty, the WTO and Changing Fundamentals of International Law.* New York: Cambridge University Press.

Jacobson, Barbara Rosen, Katharina E. Höne, and Jovan Kurbalija. 2018. *Data Diplomacy: Updating Diplomacy to the Big Data Era.* Msida, Malta: DiploFoundation.

Accessed June 11, 2021. https://www.diplomacy.edu/sites/default/files/Data_ Diplomacy_Report_2018.pdf.

Jarlner, Michael, and Thomas Østerlin Koch. 2017. "Danmark får som det første land i verden en digital ambassadør." *Politiken,* January 27, 2017. http://politiken.dk/ udland/art5806849/Danmark-f%C3%A5r-som-det-f%C3%B8rste-land-i-verden -en-digital-ambassad%C3%B8r.

Jasanoff, Sheila. 1990. *The Fifth Branch: Science Advisers as Policymakers.* Cambridge, MA: Harvard University Press.

———. 2003. "Technologies of Humility: Citizen Participation in Governing Science." *Minerva* 41(3): 223–44.

———. 2021. "Humility in the Anthropocene". *Globalizations.* https://doi.org/10.10 80/14747731.2020.1859743.

Jasanoff, Sheila, Stephen Hilgartner, Benjamin J. Hurlbut, Onur Özgöde, and Margarita Rayzberg. 2021. "Comparative Covid Response: Crisis, Knowledge, Politics. Interim Report," Harvard Kennedy School, 12 January. Accessed June 11, 2021. https://assets.website-files.com/5fdfca1c14b4b91eeaa7196a/5ffda00d50fca2e6f87 82aed_Harvard-Cornell%20Report%202020.pdf.

Jessop, Bob. 2016. *The State. Past, Present, Future.* Cambridge: Polity Press.

Jessop, Bob, and Ngai-Ling Sum. 2006. *Beyond the Regulation Approach. Putting Capitalist Economies in their Place.* Cheltenham: Edward Elgar.

Jeutner, Valentin. 2019. "The Digital Geneva Convention: A Critical Appraisal of Microsoft's Proposal." *Journal of International Humanitarian Legal Studies* 10(1): 158–70.

Jones, Michael D., Elizabeth A. Shanahan, and Mark K. McBeth. 2014. *The Science of Stories, Application of the Narrative Policy Framework in Public Policy Analysis.* New York: Palgrave Macmillan.

Jönsson, Christer, and Martin Hall. 2003. "Communication: An Essential Aspect of Diplomacy." *International Studies Perspectives* 4: 195–210.

Juncker, Jean-Claude. 2016. *State of the European Union Speech*, September 14, 2016. Accessed June 11, 2021. https://op.europa.eu/en/publication-detail/-/pub-lication/c9ff4ff6-9a81-11e6-9bca-01aa75ed71a1/language-en/format-PDF/source -30945725.

Kadlecová, Lucie, Nadia Meyer, Rafael Cos, and Pauline Ravinet. 2020. "Cyber Security: Mapping the Role of Science Diplomacy in the Cyber Field." In *Science Diplomacy in the Making: Case-based insights from the S4D4C project*, edited by M. Young, T. Flink, and E. Dall. Accessed June 11, 2021. https://www.s4d4c.eu/ wp-content/uploads/2020/03/D3.2_3_Cyber_final.pdf.

Kaltofen, Carolin, and Michele Acuto. 2018. "Science Diplomacy: Introduction to a Boundary Problem." *Global Policy* 9(S3): 8–14.

Kaltofen, Carolin, Madeline Carr, and Michele Acuto. 2019. *Technologies of International Relations. Continuity and Change.* Cham: Springer International Publishing.

Kelley, John R. 2013. "The New Diplomacy: Evolution of a Revolution." In *International Diplomacy. Volume III: The Pluralisation of Diplomacy*, edited by I. B. Neumann and H. Leira, 99–118. Thousand Oaks: Sage.

Kelsey, Jane. 2016. "TiSA: Updated Analysis of the Leaked 'Core Text' from July 2016." Accessed June 11, 2021. https://wikileaks.org/tisa/analysis/201609_TiSA_Analysis-on-Core-Text/201609_TiSA_Analysis-on-Core-Text.pdf.

Kelsey, Jane, and Burcu Kilic. 2014. "Briefing on US TiSA proposal on E-commerce, Technology Transfer, Cross-border Data Flows and Net Neutrality." *Public Services International.* Accessed June 11, 2021. http://www.world-psi.org/sites/default/files/documents/research/briefing_on_tisa_e-commerce_final.pdf.

Keohane, Robert O., and Joseph S. Nye. 2001. "The Club Model of Multilateral Cooperation and Problems of Democratic Legitimacy." In *Efficiency, Equity, and Legitimacy: the Multilateral Trading System at the Millennium,* edited by R. B. Porter, P. Sauvé, A. Subramanian, and A. Beviglia Zampetti. Washington DC: Brookings Inst. Press.

Keohane, Robert O., and Joseph S. Nye. 2011. *Power and Interdependence,* 4th ed. Stoughton: Pearson.

Kerry, Cameron, F. 2014. "Missed Connections: Talking with Europe About Data, Privacy, and Surveillance." *Center for Technology Innovation, Brookings,* May 20 2014. Accessed June 11, 2021. https://www.brookings.edu/research/missed-connections-talking-with-europe-about-data-privacy-and-surveillance/.

Kingdon, John W. 1995. *Agendas, Alternatives, and Public Policies.* New York: Longman

Kjellén, Bo. 2008. *New Diplomacy for Sustainable Development. The Challenge of Global Change.* London: Routledge.

Kleinwächter, Wolfgang. 2007. "Introduction." In *The Power of Ideas: Internet Governance in a Global Multi-Stakeholder Environment,* edited by W. Kleinwächter, 12–17. Berlin: Germany—Land of Ideas.

———. 2008. "Multistakeholderism, Civil Society, and Global Diplomacy: the Case of the World Summit on the Information Society." In *Governing Global Electronic Networks. International Perspectives on Policy and Power,* edited by W. J. Drake and E. J. Wilson III, 535–82. Cambridge: MIT Press.

———. 2017. "Internet Governance Outlook 2017: Nationalistic Heirarchies Vs. Multistakeholder Networks?" *CircleID,* January 6, 2017. Accessed June 11, 2021. https://www.circleid.com/posts/20160106_internet_outlook_2017_nationalistic_hierarchies_multistakeholder/.

Kleinwachter, Wolfgang, and Virgilio Almeida. 2015. "The Internet Governance Ecosystem and the Rainforest." *IEEE Internet Computing* 19, 64–67.

Koops, Bert-Jaap, and Rudi Bekkers. 2007. "Interceptability of Telecommunications: Is US and Dutch Law Prepared for the Future?" *Telecommunications Policy* 31: 45–67.

Kuczerawy, Aleksandra. 2015. "Intermediary Liability & Freedom of Expression: Recent Developments in the EU Notice & Action Initiative." *Computer Law & Security Review* 31(1): 46–56.

Kurbalija, Jovan. 1996. *Information Technology and Diplomacy in a Changing Environment.* Diplomatic Studies Programme. Discussion Papers(20). University of Leicester. Centre for the Study of Diplomacy.

———. 2008. "The World Summit on Information Society and the Development of Internet Diplomacy." In *Global Governance and Diplomacy. Worlds Apart?*, edited by A. F. Cooper, B. Hocking, and W. Maley, 180–207. Basingstoke: Palgrave Macmillan.

———. 2013. "The Impact of the Internet and ICT on Contemporary Diplomacy." In *Diplomacy in a Globalizing World. Theories and Practices*, edited by P. Kerr and G. Wiseman, 141–159. Oxford: Oxford University Press.

———. 2014. *An Introduction to Internet Governance*. Geneva: DiploFoundation.

———. 2016. "From Harmonising Cyberpolicies to Promoting Twiplomacy: How Diplomacy Can Strengthen Asia-Europe's Digital Connectivity." In *ASEF Outlook Report 2016/2017, CONNECTIVITY: Facts and Perspectives*. Accessed June 11, 2021. https://www.diplomacy.edu/resources/general/harmonising-cyberpolicies -promoting-twiplomacy-how-diplomacy-can-strengthen-asia.

———. 2017. *An Introduction to Internet Governance* (7 ed.). Msida, Geneva, Belgrade: DiploFoundation.

Kurbalija, Jovan, and Katharina E. Höne. 2021. "Hybrid Diplomacy: How COVID-19 Changes the Way we Negotiate." *Stifting Entwicklung und Frieden: Global Governance Spotlight* 2021(1). Accessed June 11, 2021. https://www.sef-bonn.org/en/ publications/global-governance-spotlight/12021/.

Lambright, W. Henry. 1976. *Governing Science and Technology*. New York: Oxford University Press.

Landau, Susan. 2011. *Surveillance or Security? The Risks Posed by New Wiretapping Technologies*. Cambridge: MIT Press.

Lausson, Julien. 2020. "Hadopi: une FAQ pour tout savoir." *Numerama*, April 29, 2020. Accessed June 11, 2021. https://www.numerama.com/politique/129728 -hadopi-faq-savoir.html.

Legrand, T., and Stone, Diane. 2018. "Science Diplomacy and Transnational Governance Impact." *British Politics* 13: 392–408.

Leira, Halvard. 2017. "The Making of a Classic: on Diplomacy 30 years On." *New Perspectives* 25(3): 1–7.

Lemke, Thomas. 2002. "Foucault, Governmentality, and Critique." *Rethinking Marxism* 14(3): 49–64.

Leroy, Christophe. 2016. "Existe-t-il un marché du droit public?" *Les Petites Affiches* 203:6–15.

Levinson, Nanette S. 2015. "A Tri-Decennia View of Knowledge Transfer Research: What Works in Diffusion & Development Contexts." *Journal of International Communication* 21(2): 153–68.

———. 2022. "Crafting Science Diplomacy in Comparative Perspective: The Case of U.S. Internet Governance." In *Internet Diplomacy: Shaping the Global Politics of Cyberspace*, edited by M. Marzouki and A. Calderaro. Lanham: Rowman & Littlefield.

Levinson, Nanette S., and Meryem Marzouki. 2015. "Internet Governance Institutionalization: Process and Trajectories." In *Global Governance Facing Structural Changes: New Institutional Trajectories in the Internet Era*, edited by M. Rioux and K. Fontaine-Skronski, 17–36. Palgrave-MacMillan, New York.

———. 2016. "International Organizations and Global Internet Governance Interorganizational Architecture." In *The Turn to Infrastructure in Internet Governance*, edited by F. Musiani, D. L. Cogburn, L. DeNardis, and N. S. Levinson, 47–71. New York: Palgrave Macmillan.

Lewis, James A. 2016. "Cyberspace and Armed Forces." *Australian Strategic Policy Institute*, May 1, 2016. Accessed June 11, 2021. http://www.jstor.org/stable/resrep04099.

Ley Federal. 2010. *Ley Federal de Protección de Datos Personales en Posesión de los Particulares.* DOF 05-07-2010. July 5, 2010. Accessed June 11, 2021. http://www.diputados.gob.mx/LeyesBiblio/pdf/LFPDPPP.pdf.

Lindblom Charles E. 1990. *Inquiry and Change: The Troubled Attempt to Understand and Shape Society.* New Haven: Yale University Press.

Linkov, Igor, Sankar Basu, Cathleen Fisher, Nancy Jackson, Adam C. Jones, Maija M. Kuklja, and Benjamin D. Trump. 2016. "Diplomacy for Science: Strategies to Promote International Collaboration." *Environmental Systems Decisions* 36(4): 331–34

Listek, Vanesa. 2016." Impulsan el debate sobre el Derecho al Olvido en Internet en Argentina." *La Nación.* August 30, 2016. Accessed June 11, 2021. http://www.lanacion.com.ar/1933040-impulsan-el-debate-sobre-el-derecho-al-olvido-en-internet-en-argentina/.

Lord, Kristin M., and Vaughan Turekian. 2009. "The Science of Diplomacy", *Brookings,* May 5, 2009. Accessed June 11, 2021. https://www.brookings.edu/articles/the-science-of-diplomacy/.

Loyola, Mario, and James K. Glassman. 2017. "Promoting American Values and Countering Authoritarianism in Cyberspace." *American Enterprise Institute*, February 1, 2017. Accessed June 11, 2021. https://www.aei.org/research-products/report/promoting-american-values-and-countering-authoritarianism-in-cyberspace/.

Luhmann, Niklas. 2008. *Law as a Social System.* Edited by Fatima Kastner, Richard Nobles, David Schiff, and Rosamund Ziegert. Translated by Klaus Ziegert. 1 edition. Oxford; New York: Oxford University Press.

Mabee, Bryan. 2009. *The Globalization of Security: State Power, Security Provision, and Legitimacy.* New York: Palgrave Macmillan.

Mackenzie, Donald, and Judy Wajcman. 1999. *The Social Shaping of Technology.* Buckingham: Open University Press.

MacKinnon, Rebecca, Elonnai Hickok, Allon Bar, and Hae-In, Lim. 2014. *Fostering Freedom Online: The Role of Internet Intermediaries.* Paris: UNESCO Publishing.

MacLean, Don. 2004. *Herding Schrödinger's Cats: Some Conceptual Tools for Thinking About Internet Governance.* Background Paper for the ITU Workshop on Internet Governance Geneva, 26–27 February 2004. Accessed June 11, 2021. http://www.itu.int/osg/spu/forum/intgov04/contributions/itu-workshop-feb-04-internet-governance-background.pdf.

Madiega, Tambiama. 2020. *Digital Sovereignty for Europe.* Brussels: EPRS—European Parliamentary Research Service. Accessed June 11, 2021. https://www.europarl.europa.eu/RegData/etudes/BRIE/2020/651992/EPRS_BRI(2020)651992_EN.pdf.

Manila Principles. 2015. *Manila Principles on Intermediary Liability Best Practices Guidelines for Limiting Intermediary Liability for Content to Promote Freedom of Expression and Innovation.* March 24, 2015. Accessed June 11, 2021. https://www .eff.org/files/2015/10/31/manila_principles_1.0.pdf.

Männiko, Mari. 2014. "Intermediary Service Providers' Liability Exemptions: Where Can We Draw the Line?" In *Regulating eTechnologies in the European Union*, edited by T. Kerikmäe, 119-39. Zürich: Springer International Publishing.

Manor, Ilan. 2019. *The Digitalization of Public Diplomacy.* London: Palgrave Macmillan.

Mansell, Robin. 2007. "Great Media and Communication Debates: WSIS and the MacBride Report." *Information Technologies and International Development* 4(3):15–36.

———. 2012. *Imagining the Internet: Communication, Innovation, and Governance.* Oxford: Oxford University Press.

———. 2013. "Employing Crowdsourced Information Resources: Managing the Information Commons." *International Journal of the Commons* 7(2): 255–77.

———. 2016. "Power, Hierarchy and the Internet: Why the Internet Empowers and Disempowers." *Global Studies Journal* 9(2): 19–25.

———. 2022. "Science Diplomacy and Internet Governance: Opportunities and Pitfalls." In *Internet Diplomacy: Shaping the Global Politics of Cyberspace*, edited by M. Marzouki and A. Calderaro. Lanham: Rowman & Littlefield.

Mansell, Robin, and Jean-Christophe Plantin. 2020. *Urban Futures with 5G: British Press Reporting.* London School of Economics and Political Science, June. Accessed June 11, 2021. http://eprints.lse.ac.uk/105801/.

Mansell, Robin, and W. Edward Steinmueller. 2020. *Advanced Introduction to Platform Economics.* Cheltenham: Edward Elgar Publishing.

———. 2022. "Denaturalizing Digital Platforms: Is Mass Individualization Here to Stay?" *International Journal of Communication* 16: 461–81.

Mansour, Camille. 2011. "Towards a new Palestinian Negotiation Paradigm." *Journal of Palestine Studies* 40(3):38–58.

Marchant, Eleanor, and Nicole Stremlau. 2020. "A Spectrum of Shutdowns: Reframing Internet Shutdowns from Africa." *International Journal of Communication* 14: 4327–42.

Marcinkowski, Bartosz. 2013. "Privacy Paradox(es): In Search of a Transatlantic Data Protection Standard." *Ohio State Law Journal* 74(6): 1167–92.

Marsden, Christopher T. 2017. *Network Neutrality: From Policy to Law to Regulation.* Manchester: Manchester University Press.

Marsden, Paul. 2015. "The 10 Business Models of Digital Disruption (and how to respond to them)." Last modified August 26, 2015. Accessed June 11, 2021. https://digitalwellbeing.org/the-10-business-models-of-digital-disruption-and-how -to-respond-to-them/.

Mateos Garcia, Juan, and W. Edward Steinmueller. 2008. "Open, But How Much?: Growth, Conflict, and Instititional Evolution in Open Source Communities." In *Community, Economic Creativity and Organization*, edited by A. Amin and J. Roberts, 254–81. Oxford: Oxford University Press.

Maurer, Tim. 2016. "The New Norms, Global Cyber-Security Challenges Agreements." *IHS Jane's Intelligence Review*, February 5, 2016. Accessed June 11, 2021. https://carnegieendowment.org/2016/02/05/new-norms-global-cyber-security-agreements-face-challenges-pub-63031.

Mayer-Schönberger, Viktor. 1997. "Generational Development of Data Protection in Europe." In *Technology and Privacy: The New Landscape*, edited by P. Agre and M. Rotenberg, 219–42. Cambridge: MIT Press.

Mayer-Schönberger, Viktor, and Kenneth Cukier. 2013. *Big Data: A Revolution That Will Transform How We Live, Work and Think.* Boston/NewYork: Houghton Mifflin Harcourt.

Mayer-Schönberger, Viktor, and Yann Padova. 2016. "Regime Change? Enabling Big Data through Europe's New Data Protection Regulation." *Science and Technology Law Review*, 17(2): 315–35. https://doi.org/10.7916/stlr.v17i2.4007.

Mayer, Maximilian, Mariana Carpes, and Ruth Knoblich. 2014. "The Global Politics of Science and Technology: An Introduction." In *The Global Politics of Science and Technology—Vol. 1: Concepts from International Relations and Other Disciplines*, edited by M. Mayer, M. Carpes, and R. Knoblich, 1–35. Berlin/Heidelberg: Springer.

Maynard-Moody, Steven W., and Marisa Kelly. 1993. "Stories Public Managers Tell About Elected Officials: Making Sense of the Politics-Administration Dichotomy" In *Public Management: The State of the Art*, edited by B. Bozeman, 71–90. San Francisco: Jossey-Bass.

McClory, Jonathan, and Olivia Harvey. 2016. "The Soft Power 30: Getting to Grips with the Measurement Challenge." *Global Affairs* 2(3): 309–19.

McGillivray, Kevin. 2014. "Give it Away Now? Renewal of the IANA Functions Contract and its Role in Internet Governance." *International Journal of Law and Information Technology* 22(1):3–26.

McIlwain, Charles H. 1947. *Constitutionalism: Ancient and Modern*. Indianapolis: Liberty Fund.

McKinnon, John D. 2017. "FCC Chief Ajit Pai Develops Plans to Roll Back Net Neutrality Rules", *The Wall Street Journal,* April 6, 2017. Accessed June 11, 2021. https://www.wsj.com/articles/fcc-chief-ajit-pai-develops-plans-to-roll-back-net-neutrality-rules-1491527590.

Medosch, Armin. 2001. "A Very Private Affair." *Mute* 1(21). Accessed June 11, 2021. http://www.metamute.org/editorial/articles/very-private-affair.

Mendis, Dinusha. 2013. "Digital Economy Act 2010: Fighting a Losing Battle? Why the 'Three Strikes' Law is Not the Answer to Copyright Law's Latest Challenge." *International Review of Law, Computers & Technology* 27(1-2): 60–84.

Meyer, Trisha. 2012. "Graduated Response in France: The Clash of Copyright and the Internet." *Journal of Information Policy* 2: 107–27. https://doi.org/10.5325/jinfopoli.2.2012.0107.

Meyer, Trisha, and Leo Van Audenhove. 2010. Graduated response and the emergence of a European surveillance society." *Info* 12(6): 69–79. https://doi.org/10.1108/14636691011086053.

Micciarelli, Giuseppe. 2017. "CETA, TTIP e altri fratelli. Il contratto sociale della post democrazia", in *Politica del diritto* 48(2): 231–65.

———. 2022. "Hacking the Legal. The Commons Between the Governance Paradigm and Inspirations from the 'Living History' of Collective Land Use." In *Post-Growth Planning: Towards an Urbanisation beyond the Market Economy*, edited by F. Savini, A. Ferreira, and K. C. von Schonfeld, 112–26. London: Routledge.

Milan, Stefania. 2013. *Social Movements and Their Technologies: Wiring Social Change*. Basingstoke: Palgrave Macmillan.

Milan, Stefania, and Niels ten Oever. 2017. "Coding and Encoding Rights in Internet Infrastructure." *Internet Policy Review* 6(1). https://doi.org/10.14763/2017.1.442.

Miller, Peter, and Nikolas Rose. 1990. "Governing Economic Life." *Economy and Society* 19(1), 1–31.

Moedas, Carlos. 2015. "The EU Approach to Science Diplomacy." *Remarks Delivered at the European Institute, Washington DC*, June 1, 2015.

———. 2016. "Science Diplomacy in the European Union." *Science & Diplomacy*, March 29, 2016. Accessed June 11, 2021. https://www.sciencediplomacy.org/perspective/2016/science-diplomacy-in-european-union.

Montville, Joseph. 2006. "Track Two Diplomacy: The Work of Healing History." *Whitehead Journal of Diplomacy & International Relations* 7(15): 15–25.

———. 1991a. "Transnationalism and the Role of Track-Two Diplomacy." In *Approaches to Peace; An Intellectual Map*, edited by W.S. Thompson and K.M. Jensen, 253–70. Washington: US Institute for Peace Press.

———. 1991b. "Track Two Diplomacy: The Arrow and the Olive Branch." In *The Psychodynamics of International Relations: Vol. 2, Unofficial Diplomacy at Work*, edited by V. D. Volkan, J. P. Montville, and D. A. Julius, 161–75. Lanham: Lexington.

Moore, Martin, and Damian Tambini. eds. 2018. *Digital Dominance: The Power of Google, Amazon, Facebook and Apple*. Oxford: Oxford University Press.

Moreno, Ana Elorza, Lorenzo Melchor, Gillermo Orts-Gil, Cristina Gracia, Izaskun Lacunza, Borja Izquierdo, and José Ignacio Fernández-Vera. 2017. "Spanish Science Diplomacy: a Global and Collaborative Bottom-Up Approach," Science & Diplomacy, February 14, 2017. Accessed June 11, 2021. https://www.sciencediplomacy.org/article/2017/spanish-science-diplomacy-global-and-collaborative-bottom-approach.

Morozov, Evgeny. 2017. "Moral Panic Over Fake News Hides the Real Enemy—the Digital Giants." *The Guardian*, January 8, 2017. Accessed June 11, 2021. https://www.theguardian.com/commentisfree/2017/jan/08/blaming-fake-news-not-the-answer-democracy-crisis.

Mouffe, Chantal. 2013. *Agonistics: Thinking the World Politically*. London: Verso Books.

Mueller, Milton L. 2010. *Networks and States: The Global Politics of Internet Governance*. Cambridge: MIT Press.

———. 2015. "The IANA Transition and the Role of Governments in Internet Governance." *IP Justice Journal: Internet Governance and Online Freedom Publication Series.* September 16, 2015. http://www.ipjustice.org/internet-governance/

ip-justice-journal-the-iana-transition-and-the-role-of-governments-in-internet
-governance-by-milton-mueller/.

———. 2017a. *Will the Internet Fragment? Sovereignty, Globalization, and Cyber-space*. Cambridge: Polity.

———. 2017b. "Is Cybersecurity Eating Internet Governance? Causes and Conse-quences of Alternative Framings." *Digital Policy, Regulation, and Governance* 19(6): 415–428. https://doi.org/10.1108/DPRG-05-2017-0025.

———. 2020. "Against Sovereignty in Cyberspace." *International Studies Review* 22(4): 779–801.

Mueller, Milton L., and Farzaneh Badiei. 2017. "Governing Internet Territory: ICANN, Sovereignty Claims, Property Rights and Country Code Top-Level Domains." *The Columbia Science & Technology Law Review* 18(2):435–91.

Mueller, Milton L., James Mathiason, and Hans Klein. 2007. The Internet and Global Governance: Principles and Norms for a New Regime. *Global Governance* 13(2): 237–54.

Mueller, Milton L., Andreas Schmidt, and Brenden Kuerbis. 2013. "Internet Secu-rity and Networked Governance in International Relations." *International Studies Review* 15(1): 86–104.

Musiani, Francesca. 2013. "Dangerous Liaisons? Governments, Companies and Inter-net Governance." *Internet Policy Review* 2(1). https://doi.org/10.14763/2013.1.108.

Nair, Deepak. 2019. "Saving Face in Diplomacy: A Political Sociology of Face-to-Face Interactions in the Association of Southeast Asian Nations." *European Jour-nal of International Relations* 25(3): 672–97.

Nakaschima E. 2011, "Obama Administration Outlines International Strategies for Cyberspace." *Washington Post*, May 16, 2011.

National Research Council. 1999. *The Pervasive Role of Science, Technology, and Health in Foreign Policy: Imperatives for the State*. Washington DC: The National Academies Press. https://doi.org/10.17226/9688.

———. 2015. *Diplomacy for the 21st Century: Embedding a Culture of Science and Technology Throughout the Department of State*. Washington DC: The National Academies Press. https://doi.org/10.17226/21730.

Nelson, Richard R., and Bhaven N. Sampat. 2001. "Making Sense of Institutions as a Factor Shaping Economic Performance." *Journal of Economic Behavior & Organization* 44(1): 31–54.

NETmundial. 2014. *NETmundial Multistakeholder Statement*. April 24, 2014. Accessed June 11, 2021. http://netmundial.br/wp-content/uploads/2014/04/NET-mundial-Multistakeholder-Document.pdf.

Neumann, Iver B. 2008. Diplomacy and Globalisation. In *Global Governance and Diplomacy. Worlds Apart?*, edited by A. F. Cooper, B. Hocking, and W. Maley 15–28. Basingstoke: Palgrave Macmillan.

———. 2011. *At Home with the Diplomats: Inside a European Foreign Ministry*. Ithaca: Cornell University Press.

Newman, Nathan. 2014. "Search, Antitrust and the Economics of the Control of User Data." *Yale Journal on Regulation* 31(2): 401–54.

Ney, Steven. 2006. "Messy Issues, Policy Conflict and the Differentiated Polity: Ana-lysing Contemporary Policy Responses to Complex, Uncertain and Transversal Policy Problems." Phd diss., Vienna, LOS Center for Bergen.

Nye, Joseph S. 1990. "Soft Power." *Foreign Policy, 80*(Autumn): 153–71.

———. 2004. *Soft Power: The Means to Success in World Politics*. New York: PublicAffairs.

———. 2008. "Public Diplomacy and Soft Power." *Annals of the American Academy of Political and Social Science* 616(1): 94–109.

———. 2014. *The Regime Complex for Managing Global Cyber Activities.* Global Commission on Internet Governance Paper Series N°1. https://www.cigionline .org/sites/default/files/gcig_paper_no1.pdf.

———. 2017. "A Normative Approach to Preventing Cyberwarfare." *AZERNEWS*, March 15, 2017. Accessed June 11, 2021. http://www.azernews.az/analysis/110166. html.

OECD. 1980. *OECD Guidelines on the Protection of Privacy and Transborder Flows of Personal Data*. Accessed June 11, 2021. http://www.oecd.org/sti/ieconomy/ oecdguidelinesontheprotectionofprivacyandtransborderflowsofpersonaldata.htm.

———. 2012. *Cybersecurity Policy Making at a Turning Point: Analysing a New generation of National Cybersecurity Strategies for the Internet Economy*. Paris: OECD. Accessed June 11, 2021. https://www.oecd.org/sti/ieconomy/cybersecu-rity%20policy%20making.pdf.

———. 2013. *OECD Revised Guidelines on the Protection of Privacy and Trans-border Flows of Personal Data*. Accessed June 11, 2021. http://www.oecd.org/ internet/ieconomy/privacy-guidelines.htm.

Orefice, Maria. 2016. "Big Data. Regole e concorrenza." *Politica del diritto* 47(4): 697–743.

Ostrom, Elinor. 1990. *Governing the Commons: The Evolution of Institutions for Col-lective Action*. Cambridge: Cambridge University Press.

Owen, John M. 2012. "Graduated Response Systems and the Market for Copyrighted Works." *Berkeley Technology Law Journal* 17(4): 557–612.

Palladino, Nicola, and Mauro Santaniello. 2021. Legitimacy, Power, and Inequali-ties in the Multistakeholder Internet Governance: Analyzing the IANA Transition. London: Palgrave Macmillan.

Parkinson, C. Northcote. 1958. *The Evolution of Political Thought*. Boston: Houghton Mifflin.

Pauwels, Caroline, and Simon Delaere. 2007. "The Political and Regulatory Frame-work towards a European Information and Knowledge Society." In *The Privatisa-tion of European Telecommunications*, edited by K. Eliassen and J. From, 51–74. Burlington: Ashgate Publishing Company.

Penca, Jerneja. 2018. "The Rhetoric of "science Diplomacy": Innovation for the EU's Scientific Cooperation?" *EL-CSID Working Paper Issue 2018/16,* April 2018. https://doi.org/10.5281/zenodo.1228016.

Peng, ShinYi. 2013. "Is the Trade in Services Agreement (TiSA) a Stepping Stone for the Next Version of GATS?" *Hong Kong Law Journal* 43(2): 611–32

Pérez, Ana Lilia. 2007. "Fraude en Estrella Blanca alcanza a Vamos México". *Fortuna* no. 49, February 2007.

Peters, Claire. 2022. "National Sovereignty, Global Policy, and the Liberalization of Telecommunications Markets." In *Internet Diplomacy: Shaping the Global Politics of Cyberspace*, edited by M. Marzouki and A. Calderaro. Lanham: Rowman & Littlefield.

Petiteville, Franck. 2018. "International Organizations Beyond Depoliticized Governance." *Globalizations* 15(3): 301–13.

Pezziardi, Pierre, and Henri Verdier. 2017. *Des startups d'État à l'État plateforme*. Paris: Fondation pour l'innovation politique.

Pigman, Geoffrey A. 2010. *Contemporary Diplomacy: Representation and Communication in a Globalized World*. Cambridge: Polity.

Piper, Fred, and Michael Walker. 1998. "Cryptographic Solutions for Voice Telephony and GSM." *Network Security* 1998(12): 14–19.

Pohle, Julia. 2015. "Multistakeholderism Unmasked: How the NetMundial Initiative Shifts Battlegrounds in Internet Governance." *Global Policy*, January 5, 2015. https://www.globalpolicyjournal.com/blog/05/01/2015/multistakeholderism -unmasked-how-netmundial-initiative-shifts-battlegrounds.

———. 2016. "Multistakeholder Governance Processes as Production Sites: Enhanced Cooperation 'In the Making'." *Internet Policy Review* 5(3). https://doi .org/10.14763/2016.3.432.

Pollach, Irene. 2005. "A Typology of Communicative Strategies in Online Privacy Policies: Ethics, Power and Informed Consent." *Journal of Business Ethics* 62: 221–35.

Post, David G., and Danielle Kehl. 2015. "Controlling Internet Infrastructure, Part 1: The IANA Transition and Why It Matters for the Future of the Internet." Open Technology Institute, New America Foundation, April 30, 2015. https://www .newamerica.org/oti/policy-papers/controlling-internet-infrastructure-part-i/.

Poteete, Amy R., Marco A Janssen, and Elinor Ostrom. 2010. *Working Together: Collective Action, the Commons, and Multiple Methods in Practice*. Princeton: Princeton University Press.

Pouliot, Vincent. 2011. "Diplomats as Permanent Representatives. The Practical Logics of the Multilateral Pecking Order." *International Journal* 66(3): 543–61.

———. 2016. "Hierarchy in Practice: Multilateral Diplomacy and the Governance of International Security." *European Journal of International Security* 1(1): 5–26.

Pouliot, Vincent, and Jérémie Cornut. 2015. "Practice Theory and the Study of Diplomacy: A Research Agenda." *Cooperation and Conflict* 50(3): 297–315. https://doi .org/10.1177/0010836715574913.

Powell, Alison B. 2015. "Open Culture and Innovation: Integrating Knowledge Across Boundaries." *Media, Culture and Society* 37(3): 376–93.

Powers, Shawn M., and Michael Jablonski. 2015. *The Real Cyber War: The Political Economy of Internet Freedom*. Urbana: Illinois University Press.

Presidência da República. 2014. *Marco Civil da Internet—The Brazilian Civil Framework of the Internet.* translation by the Brazilian Chamber of Deputies. Lei 12.965.

April 23, 2014. Accessed June 11, 2021. http://bd.camara.gov.br/bd/bitstream/ handle/bdcamara/26819/bazilian_framework_%20internet.pdf.

La Quadrature du Net. 2009. "Net Freedoms in Europe: Recapitulating the Capitulation." *La Quadrature du Net*, October 31, 2009. Accessed June 11, 2021. http:// www.laquadrature.net/en/net-freedoms-in-europe-recapitulating-the-capitulation.

R3D. 2016. "¡Ganamos! Tribunal anula resolución del INAI sobre el falso "derecho al olvido". R3D website. Accessed June 11, 2021. https://r3d.mx/2016/08/24/amparo -inai-derecho-olvido/.

Radu, Roxana. 2019. *Negotiating Internet Governance*. Oxford: Oxford University Press.

Radu, Roxana, and Jean-Marie Chenou. 2014. "Global Internet Policy. A Fifteen-Year long Debate." In *The Evolution of Global Internet Governance. Principles and Policies in the Making*, edited by R. Radu, J.-M. Chenou, and R. H. Weber, 1–19. Berlin: Springer.

Ramel, Frédéric, and Cécile Prévost-Thomas, eds. 2018. *International Relations, Music and Diplomacy. Sounds and Voices on the International Stage*. London: Palgrave Macmillan.

Ranaivoson, Heritiana, and Anne-Catherine Lorrain. 2012. "Graduated Response Beyond the Copyright Balance: Why and How the French HADOPI Takes Consumers as Targets." *Info* 14(6): 34–44.

Rathbun, Brian C. 2014. *Diplomacy's Value. Creating Security in 1920s Europe and the Contemporary Middle East*. Ithaca: Cornell University Press.

Raymond, Mark, and Laura DeNardis. 2015. "Multistakeholderism: Anatomy of an Inchoate Global Institution." *International Theory* 7(03): 572–616.

Renner, Morritz. 2010. *Zwingendes transnationales Recht: Zur Struktur einer Wirtschaftverfassung jenseits des Staates*. Baden-Baden: Nomos.

Rhoads, Christopher, and Loretta Chao. 2009. "Iran's Web Spying Aided by Western Technology." *Wall Street Journal*, June 22, 2009. Accessed June 11, 2021. http:// www.wsj.com/articles/SB124562668777335653.

Riordan, Jaani. 2016. *The Liability of Internet Intermediaries*. Oxford: Oxford University Press.

Riordan, Shaun. 2019. *Cyberdiplomacy, Managing Security and Governance Online*. Cambridge: Polity Press.

Robin, Ron. 2005. "Requiem for Public Diplomacy?" *American Quarterly* 57(2): 345–53.

Rodine-Hardy, Kirsten. 2013. *Global Markets and Government Regulation in Telecommunications*. New York: Cambridge University Press.

Rodríguez-Garavito, César A. 2012. "Towards a Sociology of the Rule of Law Field: Neoliberalism, Neoconstitutionalism, and the Contest over Judicial Reform in Latin America." In *Lawyers and the Rule of Law in an Era of Globalization*, edited by Y. Dezalay and B. Garth, 156–82. London: Routledge.

Rogers, Michael, and Arlene Luck. 2017. "Digital Citizenship and Surveillance: The Snowden Disclosures, Technical Standards, and the Making of Surveillance Infrastructures." *International Journal of Communication* 11: 802–23.

The Royal Society. 2010. *New Frontiers in Science Diplomacy: Navigating the Changing Balance of Power*. London: The Royal Society. https://royalsociety.org/~/media/Royal_Society_Content/policy/publications/2010/4294969468.pdf.

———. 2012. *Science as an Open Enterprise*. London: The Royal Society.

Rozgonyi, Krisztina, and Katharine Sarikakis. 2022. "Policy diffusion and Internet governance: reflections on copyright and privacy." In *Internet Diplomacy: Shaping the Global Politics of Cyberspace*, edited by M. Marzouki and A. Calderaro. Lanham: Rowman & Littlefield.

Rüffin, Nicolas. 2020. "EU Science Diplomacy in a Contested Space of Multi-level Governance: Ambitions, Constraints and Options for Action." *Research Policy* 49(1): 1–10.

Ruohonen, Jukka, Sami Hyrynsalmi, and Ville Leppänen. 2016. "An Outlook on the Evolution of the European Union Cyber Security Apparatus." *Government Information Quarterly 33(*4): 746–56.

S4D4C. 2019. Madrid Declaration on Science Diplomacy. Accessed June 11, 2021. https://www.s4d4c.eu/s4d4c-1st-global-meeting/the-madrid-declaration-on-science-diplomacy/.

Sabatier, Paul A. 1993. "Policy Change over a Decade or More." In *Policy Change and Learning: An Advocacy Coalition Approach*, edited by P. A. Sabatier and H. Jenkins-Smith, 13–39. Boulder: Westview Press.

Saldías, Osvaldo. 2012. *Patterns of Legalization in the Internet: Do We Need a Constitutional Theory for Internet Law?* HIIG Discussion Paper Series n°2012 -08. Accessed June 11, 2021. https://papers.ssrn.com/sol3/papers.cfm?abstract_id=1942161.

Sanger, David E., and John Markoff 2009. "Obama Outlines Coordinated Cybersecurity Plan." *New York Times*, May 30, 2009.

Santaniello, Mauro, and Nicola Palladino. 2022. "Discourse Coalitions in *Internet Governance: Shaping Global Policy by Narratives and Definitions.*" In *Internet Diplomacy: Shaping the Global Politics of Cyberspace*, edited by M. Marzouki and A. Calderaro. Lanham: Rowman & Littlefield.

Sargsyan, Tatevik. 2016. "The Privacy Role of Information Intermediaries through Self-Regulation." *Internet Policy Review* 5(4). Https://doi.org/10.14763/2016.4.438.

Sarikakis, Katharine. 2004. *Powers in Media Policy. The Challenge of the European Parliament*. Oxford/Bern: Peter Lang.

———. 2012a. "Securitization and Legitimacy in Global Media Governance." In *The Handbook of Global Media Research*, edited by I. Volkmer, 143–55. Oxford: Wiley-Blackwell.

———. 2012b. "Serving Two Masters: The Roles of the Market and European Politics in the Governance of Media Transformations." In *Understanding Media Policies*, edited by E. Psychogiopoulou, 247–56. Basingstoke: Palgrave Macmillan.

———. 2015. "The Struggle for Control Over Communicative Spaces: Creating, Sustaining, Resisting as Tactics of Information." *IS4IS Summit Vienna 2015*. Accessed June 11, 2021. https://sciforum.net/manuscripts/2988/manuscript.pdf.

Sarikakis, Katharine, and Sarah Gantner. 2014. "Priorities in Global Media Policy Transfer: Audiovisual and Digital Policy Mutations in the EU, MERCOSUR and the U.S. triangle." *European Journal of Communication* 29(1): 17–33.

Sarikakis, Katharine, and Joan R. Rodriguez-Amat. 2013. Copyright and Privacy Governance: Policy Intersections and Challenges." In *Communication and Media Policy in the Era of the Internet. Theories and Processes*, edited by M. Löblich and S. Pfaff-Rüdiger, 147–158. Baden-Baden: Nomos.

Sartori, Giovanni. 2000. *Elementi di Teoria Politica*. Bologna: Il Mulino.

Sassen, Saskia. 2007. *A Sociology of Globalization*. New York-London: W.W. Norton & Company Inc.

Sauvé, Pierre. 2014. "A Plurilateral Agenda for Services? Assessing the Case for a Trade in Services Agreement (TiSA)." In *The Preferential Liberalization of Trade Services*, edited by P. Sauvé and A. Shingal. Cheltenham-Northampon: Edward Elgar Publishing.

Schemeil, Yves. 2004. "Expertise and Political Competence: Consensus Making within the World Trade and the World Meteorological Organizations." In *Decision-Making Within International Organizations*, edited by B. Reinalda and B. Verbeek, 77–89. London: Routledge.

Schemeil, Yves. 2012. "Global Governance of the Information System Revisited: Evolution or Innovation in International Politics?" In *Governance, Regulations and Powers on the Internet*, edited by E. Brousseau, M. Marzouki and C. Meadel, 186–208. Cambridge: Cambridge University Press.

———. 2013. "Bringing International Organization In: Global Institutions as Adaptive Hybrids." *Organization Studies* 34(2): 219–52.

———. 2022. "Undiplomatic Ties: When Internet Blocks Intermediation." In *Internet Diplomacy: Shaping the Global Politics of Cyberspace*, edited by M. Marzouki and A. Calderaro. Lanham: Rowman & Littlefield.

Schmid, Gerhard. 2001. *Report on the Existence of a Global System for the Interception of Private and Commercial Communications (ECHELON interception system)*. Brussels/Strasbourg: European Parliament. Accessed June 11, 2021. http://www.europarl.europa.eu/sides/getDoc.do?pubRef=-//EP//TEXT+REPORT+A5-2001-0264+0+DOC+XML+V0//EN.

Schmitt, Carl. 1932. *Das Zeitalter der Neutralisierungen und Entpolitisierungen. Europäische Revue.* Italian translation, "L'epoca delle neutralizzazioni e delle spoliticizzazioni", in *Le categorie del «politico»*. Bologna: il Mulino, 1972.

Schneider, Volker. 2002. "The Institutional Transformation of Telecommunications between Europeanization and Globalization." In *Governing Telecommunications and the New Information Society in Europe*, edited by J. Jordana, 27–46. Cheltenham: Edward Elgar Publishing.

Scholte, Jan A. 2002. "Civil Society and Democracy in Global Governance." *Global Governance* 8(3): 281–304.

———. 2005. *Globalization: A Critical Introduction*. 2nd Rev Edition. New York: Red Globe Press.

———. 2014. "Reinventing Global Democracy." *European Journal of International Relations* 20(1): 3–28.

————. 2020. *Multistakeholderism Filling the Global Governance Gap?* Stockholm: Global Challenges Foundation. Accessed June 11, 2021. https://globalchallenges .org/wp-content/uploads/Research-review-global-multistakeholderism-scholte -2020.04.06.pdf.

Schrempf, Judith. 2011. "Nokia Siemens Networks: Just Doing Business—or Supporting an Oppressive Regime?" *Journal of Business Ethics* 103(1): 95–110.

Schwartz, Paul M. 2002. "Privacy, Participation and Cyberspace." In *Perspektive Datenschutz. Praxis und Entwicklungen in Recht und Technik*, edited by B. Baeriswyl, and B. Rudin, 67–85. Zürich: Schulthess Juristische Medien AG.

Selleslaghs, Joren. 2017. "EU-Latin American Science Diplomacy." *EL-CSID Working Paper Issue 2017/8*, September 2017.Accessed June 11, 2021. https://www .ies.be/files/EL-CSID_WorkingPaper_2017-08_EU-Latin_AmericanScienceDiplomacy.pdf.

Sending, Ole Jacob, Jérémie Pouliot, and Iver B. Neuman. 2011. "The Future of Diplomacy. Changing Practices, Evolving Relationships." *International Journal* 66(3): 527–42.

Shahbaz, Adrian, and Allie Funk. 2020. *Freedom on the Net 2020: The Pandemic's Digital Shadow.* Washingdon DC: Freedom House. Accessed June 11, 2021. https://freedomhouse.org/sites/default/files/2020-10/10122020_FOTN2020_Complete_Report_FINAL.pdf

Shanahan, Elizabeth A., Stephanie M. Adams, Michael D. Jones, and Mark K. McBeth. 2014. "The Blame Game: Narrative Persuasiviness of the Intentional Causal Mechanism." In *The Science of Stories, Application of the Narrative Policy Framework in Public Policy Analysis*, edited by M. D. Jones, E. A. Shanahan, and M. K. McBeth, 69–88. New York: Palgrave Macmillan.

Shanahan, Elizabeth A., Mark K. McBeth, and Michael D. Jones. 2011. "Policy Narrative and Policy Process." *Policy Studies Journal* 39(4): 535–61.

Sharp, Paul. 1997. "Who Needs Diplomats? The Problem of Diplomatic Representation." *International Journal* 52(4): 609–34.

————. 1999. "For Diplomacy: Representation and the Study of International Relations." *International Studies Review* 1(1): 33–57.

————. 2004. "The Idea of Diplomatic Culture and its Sources." In *Intercultural Communication and Diplomacy*, edited by H. Slavik, 361–79. Malta/Geneva: DiploFoundation.

————. 2019. *Diplomacy in the 21st Century. A Brief Introduction.* Abingdon, Oxon: Routledge.

Simmons, Beth A., Franck Dobbin, and Geoffrey Garrett. 2006. "Introduction: The International Diffusion of Liberalism." *International Organization* 60(4): 781–810.

Singh, Harsha Vardhana, Ahmed Abdel-Latif, and L. Lee Tuthill. 2016. "Governance of International Trade and the Internet: Existing and Evolving Regulatory Systems." *Global Commission on Internet Governance,* Paper Series: No. 32. Accessed June 11, 2021. https://www.cigionline.org/sites/default/files/gcig_no32web_0.pdf.

Slaughter, Anne-Marie. 2004. *A New World Order.* Princeton: Princeton University Press.

Śledziewska, Katarzyna, and Renata Wloch. 2017. "Should We Treat Big Data as a Public Good?" In *The Responsibilities of Online Service Providers*, edited by M. Taddeo and L. Floridi, 263–73. Cham ZG: Springer.

Soto Galindo, José. 2016. "Fortuna obliga al INAI a discutir sobre Google y los datos personales otra vez." *El Economista* August 25, 2016. Accessed June 11, 2021. http://eleconomista.com.mx/economicon/2016/08/25/fortuna-obliga-inai-discutir-sobre-google-datos-personales-otra-vez.

Spillman, Lyn. 2017. "Meta-Organizations Matter." *Journal of Management Inquiry* 27(1): 16–20.

Stiglitz, Joseph E. 2002. *Globalization and Its Discontents*. New York: W.W. Norton Company.

Stodden, Victoria. 2010. "Open Science: Policy Implications for the Evolving Phenomenon of User-Led Scientific Innovation." *Journal of Science Communication* 9(1): 1–8.

Stone, Deborah A. 1988. *Policy Paradox and Political Reason*. Glenview: Scott, Foresman and Company.

———. 2002. *Policy Paradox: The Art of Political Decision Making*. New York: Norton.

Stone, Diane. 2019. *Making Global Policy (Elements in Public Policy)*. Cambridge: Cambridge University Press.

Strange, Susan. 1998. *States and Markets*. Second edition. London: Continuum.

Superior Tribunal de Justiça. 2016. *SMS v. Google Brasil Internet Ltda. Nº 1.593.873—SP (2016/0079618-1)*. Accessed June 11, 2021. http://www.internetlab.org.br/wp-content/uploads/2017/02/STJ-REsp-1.593.873.pdf.

Sutherland, William J., Laura Bellingan, Jim R. Bellingham, Jason J Blackstock, Robert M Bloomfield, Michael Bravo, Victoria M. Cadman, et al. 2012. "A Collaboratively-Derived Science-Policy Research Agenda." *PLoS ONE* 7(3): 1–5.

Suzor, Nicolas, and Brian Fitzgerald. 2011. "The Legitimacy of Graduated Response Schemes in Copyright Law." *University of New South Wales Law Journal* 43(1): 1–40.

Tanczer, Leonie M., Irina Brass, and Madeline Carr. 2018. "CSIRTs and Global Cybersecurity: How Technical Experts Support Science Diplomacy." *Global Policy* 9(S3): 60–66.

Tang, Min. 2020. "Huawei Versus the United States? The Geopolitics of Extraterritorial Internet Infrastructure." *International Journal of Communication* 14: 4556–77.

Taplin, Jonathan. 2017. *Move Fast and Break Things: How Facebook, Google, and Amazon Cornered Culture and Undermined Democracy*. New York: Little, Brown and Company.

Teubner, Günther. 1993. *Law as an Autopoietic System*. Oxford: Blackwell.

———. 2011. "Transnational Fundamental Rights: Horizontal Effect?" *Netherlands Journal of Legal Philosophy* 3: 191–215.

———. 2012. *Constitutional Fragments: Societal Constitutionalism and Globalization*. Oxford: Oxford University Press.

Theoharis, Athan. 1992. "FBI Wiretapping: A Case Study of Bureaucratic Autonomy." *Political Science Quarterly* 107(1): 101–122.

Thomson, Janice E. 1995. "State Sovereignty in International Relations: Bridging the Gap Between Theory and Empirical Research." *International Studies Quarterly* 39(2): 213–33.

Trager, Robert F. 2017. *Diplomacy. Communication and the Origins of International Order.* Cambridge: Cambridge University Press.

Traynor, Ian. 2014. "Internet Governance Too US-Centric, Says European Commission." *The Guardian* February 12, 2014. Accessed June 11, 2021. https://www.theguardian.com/technology/2014/feb/12/internet-governance-us-european-commission.

Tucci, Antonio. 2012. *Immagini del diritto. Tra fattualità istituzionalistica e agency.* Torino: Giappichelli.

Turekian, Vaughan C. 2012. "Building a National Science Diplomacy System." *Science & Diplomacy*, December 10, 2012. Accessed June 11, 2021. https://www.sciencediplomacy.org/editorial/2012/building-national-science-diplomacy-system.

Turekian, Vaughan C., Sarah Macindoe, Daryl Copeland, Lloyd S. Davis, Robert G. Patman, and Maria Pozza. 2015. "The Emergence of Science Diplomacy." In *Science Diplomacy: New Day or False Dawn?*, edited by L. S. Davis and R. G. Patman, 3–24. London: World Scientific Publishing.

Turekian, Vaughan C., Gluckman, Peter D, Teruo Kishi, and Robin W Grimes. 2018. "Science Diplomacy: A Pragmatic Perspective from the Inside." *Science & Diplomacy*, January 16, 2018. Accessed June 11, 2021. https://www.sciencediplomacy.org/article/2018/pragmatic-perspective.

Ülgen, Sinan. 2016. *Governing Cyberspace: A Road Map for Transatlantic Leadership,* Carnegie Europe. Accessed June 11, 2021. https://carnegieendowment.org/files/Sinan_Cyber_Final.pdf.

Ulnicane, Inga, Damian Okaibedi Eke, William Knight, George Ogoh, and Bernd Carsten Stahl. 2021. Good Governance as a Response to Discontents? Déjà Vu, or Lessons for AI from Other Emerging Technologies." *Interdisciplinary Science Reviews*, 46(1–2): 71–93.

UN. 2018. SG/A/1817 *Secretary-General Appoints High-Level Panel on Digital Cooperation.* New York: UN.

———. 2019. *The Age of Digital Interdependence.* High-Level Panel on Digital Cooperation. Accessed June 11, 2021. https://www.un.org/en/pdfs/DigitalCooperation-report-for%20web.pdf.

———. 2020. *United Nations Strategy and Plan of Action on Hate Speech: Detailed guidance on Implementation for United Nations Field Presences.* New York: United Nations. Accessed June 11, 2021. https://www.un.org/en/genocideprevention/documents/UN%20Strategy%20and%20PoA%20on%20Hate%20Speech_Guidance%20on%20Addressing%20in%20field.pdf.

United Nations Conference on Trade and Development (UNCTAD). 2016. *Data Protection Regulations and International Data Flows: Implications for Trade and Development.* New York/Geneva: United Nations. Accessed June 11, 2021. http://unctad.org/en/PublicationsLibrary/dtlstict2016d1_en.pdf.

UNESCO. 2021. Communication and Information Programme. Paris: UNESCO. Accessed June 11, 2021. https://en.unesco.org/ci-programme.

UNGA. 2018. A/RES/73/27 *Developments in the Field of Information and Telecommunications in the Context of International Security.* New York: UN.

———. 2020. A/74/821 *Road Map for Digital Cooperation: Implementation of the Recommendations of the High-Level Panel on Digital Cooperation.* New York: UN.

Ureste, Manu. 2016. "Derecho al olvido en internet: ¿un derecho, censura o un redituable negocio en México?" *Animal Político*, September 13, 2016. Accessed June 11, 2021. http://www.animalpolitico.com/2016/09/derecho-olvido-internet -censura-mexico/.

US DoC. 2000. Issuance of Safe Harbor Principles and Transmission to European Commission, by United States Department of Commerce, July 24, 2000. Accessed June 11, 2021. https://www.federalregister.gov/documents/2000/07/24/00-18489/ issuance-of-safe-harbor-principles-and-transmission-to-european-commission.

V.J. 2016. "Téléchargement illégal: Hadopi a doublé le nombre de dossiers transmis à la justice en un an." *20 Minutes*, July 28, 2016. Accessed June 11, 2021. http:// www.20minutes.fr/high-tech/1901119-20160728-telechargement-illegal-hadopi -double-nombre-dossiers-transmis-justice-an.

Van Dijck, José, Thomas Poell, and Martijn de Waal. 2018. *The Platform Society: Public Values in a Connective World.* Oxford: Oxford University Press.

Van Eeten, Michel. J. G., and Milton. L. Mueller. 2013. "Where is the Governance in Internet Governance?" *New Media & Society* 15(5): 720–36.

Van Langenhove, Luk. 2016a. "Global Science Diplomacy as a New Tool for Global Governance." *Pensament* n°3. Barcelona: FOCIR. Accessed June 11, 2021. https:// cris.unu.edu/global-science-diplomacy-new-tool-global-governance.

———. 2016b. "Global Science Diplomacy for Multilateralism 2.0." *Science & Diplomacy*, December 29, 2016. Accessed June 11, 2021. https://www.sciencedi-plomacy.org/article/2016/global-science-diplomacy-for-multilateralism-20.

———. 2017. "Tool for an EU Science Diplomacy." Luxembourg: Publication Office of the EU. https://doi.org/10.2777/911223.

———. 2019. "Who cares? Science Diplomacy and the Global Commons." *Australian Quarterly* 90(4): 18–27.

Vargas, Paula. 2016. *El derecho a remover contenido de Internet: ¿que límites impone el sistema interamericano de protección de la libertad de expresión?* Centro de Estudios en Libertad de Expresión y Acceso a la Información, Universidad de Palermo. June 2016. Accessed June 11, 2021. http://responsible-tech.org/wp -content/uploads/2016/06/Derecho-al-Olvido-1.pdf.

Vatu, Gabriela. 2014. "Artist Protesting Hadopi Law Is Censored in Paris. The Anti-Piracy Law Isn't Anyone's Favorite in France." *Softpedia News*, February 17, 2014. Accessed June 11, 2021. http://news.softpedia.com/news/Artist-Protesting -Hadopi-Law-Is-Censored-by-Paris-427299.shtml.

Vauchez, Antoine. 2015. *Brokering Europe: Euro-Lawyers and the Making of a Transnational Polity.* Cambridge: Cambridge University Press.

Vercellone, Carlo, Francesca Bria, Andrea Fumagalli, Eleonora Gentilucci, Alfonso Giuliani, Giorgio Griziotti, and Pierluigi Vattimo. 2015. "Managing the Commons in the Knowledge Economy." Report of the D-Cent Project. Accessed June 11, 2021. https://halshs.archives-ouvertes.fr/halshs-01180341/document.

Verweij, Marco, and Thompson, Michael, eds. 2006. *Clumsy Solutions for a Complex World: Governance, Politics, and Plural Perceptions*. Houndmills: Palgrave Macmillan.

Vijayan, Jaykumar. 2010. "U.S. Should Seek World Cooperation on Cyber Conflict, Says ex-CIA Director." Computerworld.com, July 29, 2010.

Visco Comandini Vincenzo. 2018. "Le fake news sui social network: un'analisi economica." *Medialaws—Rivista di diritto dei media* 2018(2): 183–212.

Vixie, Paul. 2010. "Taking Back the DNS", *CircleID,* July 20, 2010. Accessed June 11, 2021. http://www.circleid.com/posts/20100728_taking_back_the_dns/.

Vixie, Paul, and Vernon Schryver. 2017. "DNS Response Policy Zones (RPZ), Draft 00". *IETF*, Domain Name System Operations. Accessed June 11, 2021. https://tools.ietf.org/html/draft-ietf-dnsop-dns-rpz-00.

Vogel, Kenneth, P. 2017. "Google Critic Ousted From Think Tank Funded by the Tech Giant." *New York Times*, August 30, 2017. Accessed June 11, 2021. https://www.nytimes.com/2017/08/30/us/politics/eric-schmidt-google-new-america.html?mcubz=3.

Warren, Mark. 1989. "On Diplomacy: A Genealogy of Western Estrangement. by James Der Derian, Review." *The Journal of Politics* 51(1): 208–11.

Watson, Ivan. 2010. "Nokia Siemens Says It Didn't Help Iranian Government Spy." CNN, August 20, 2010. Accessed June 11, 2021. http://edition.cnn.com/2010/WORLD/meast/08/20/iran.us.nokia.siemans.lawsuit/.

Weber, Rolf H. 2012. "Regulatory Autonomy and Privacy Standards under the GATS." *Asian Journal of WTO & International Health Law and Policy* 7(1): 25–48.

Wedlin, Linda, and Maria Nedeva. 2015. "Towards European Science: An Introduction." In *Towards European Science—Dynamics and Policy of an Evolving European Research Space*, edited by L. Wedlin and M. Nedeva, 1-11. Cheltenham: Edward Elgar Publishing.

Wendt, Alexander. 1992. "Anarchy Is What States Make of It: The Social Construction of Power Politics." *International Organization* 46(2): 391–425. doi:10.2307/2706858.

———. 1999. *Social Theory of International Politics*. Cambridge: Cambridge University Press.

WGIG. 2005. *Report of the Working Group on Internet Governance*. Geneva: ITU. http://www.itu.int/net/wsis/wgig/docs/wgig-report.pdf.

Wiener, Jonathan B., and Alberto Alemanno. 2015. "The Future of International Regulatory Cooperation: TTTIP as a Learning Process Toward a Global Policy Laboratory." *Law and Contemporary Problems* 78(4): 103–36.

Wight, Colin. 2006. *Agents, Structures, and International Relations: Politics as Ontology*. Cambridge: Cambridge University Press.

Wight, Martin, and Herbert Butterfield, eds. 1969. *Diplomatic Investigations. Essays in the Theory of International Politics.* London: Allen & Unwin.

WikiLeaks. 2014. *Secret TPP treaty: Intellectual Property Chapter working document for all 12 nations with negotiating positions.* October, 16, 2014. Accessed June 11, 2021. https://www.wikileaks.org/tpp-ip2/tpp-ip2-chapter.pdf.

Williams, Raymond. 1983. *Towards 2000.* London: The Hogarth Press.

Wiseman, Geoffrey. 2004. "'Polylateralism' and New Modes of Global Dialogue." In *Diplomacy (Volume III): Problems and Issues in Contemporary Diplomacy*, edited by C. Jönsson and R. Langhorne, 36–57. Thousand Oaks: Sage.

———. 2011. "Norms and Diplomacy: the Diplomatic Underpinnings of Multilateralism." In *The New Dynamics of Multilateralism: Diplomacy, International Organizations, and Global Governance*, edited by J. P. Muldoon, J. F. Aviel, R. Reitano, and E. Sullivan, 5–22. Boulder: Westview Press.

———. 2015. "Diplomatic Practices at the United Nations." *Cooperation and Conflict* 50(3): 316–33.

Witjes, Nina. 2017. "The Co-Production of Science, Technology and Global Politics: Exploring Emergent Fields of Knowledge and Policy." PhD Diss., Technical University Munich. Accessed June 11, 2021. https://mediatum.ub.tum.de/doc/1350479/1350479.pdf.

Wong, Ken. 1995. "Fighting Mobile Phone Fraud—Who is Winning?—Part 1." *Computer Fraud & Security Bulletin* 1995(1): 9–16.

Wong, Seanon S. 2016. "Emotions and the Communication of Intentions in Face-to-Face Diplomacy." *European Journal of International Relations* 22(1): 144–67.

Yang, Aimei, Rong Wang, and Jian (Jay) Wang. 2017. "Green Public Diplomacy and Global Governance: The Evolution of the US–China Climate Collaboration Network, 2008–2014." *Public Relations Review* 43(5): 1048–61

Yu, Peter K. 2010. "The Graduated Response." *Florida Law Review* 62:1373–1430.

Zimmermann, Phil. 1999. "Why I Wrote PGP." Accessed June 11, 2021. https://www.philzimmermann.com/EN/essays/WhyIWrotePGP.html.

Zittoun, Philippe. 2014. *The Political Process of Policymaking.* New York: Palgrave Macmillan.

Zittrain, Jonathan. 2006. "A History of Online Gatekeeping." *Harvard Journal of Law & Technology* 19(2): 253–300.

Zuboff, Shoshana. 2019. *The Age of Surveillance Capitalism: The Fight for a Human Future at the New Frontier of Power.* New York: Public Affairs.

Zürn, Michael. 2018. *A Theory of Global Governance: Authority, Legitimacy, and Contestation.* Oxford: Oxford University Press.

Zysman, John, and Steven Weber. 2001. "Governance and Politics of the Internet Economy: Historical Transformation of Ordinary Politics with a New Vocabulary?" BRIE Working Paper 141, BRIE, University of California, Berkeley. Accessed June 11, 2021. https://brie.berkeley.edu/sites/default/files/wp141.pdf.

Index

4S. *See* Society for the Social Studies of Science

A2IM. *See* American Association of Independent Music

AAAS. *See* American Association for the Advancement of Science

access:
freedom to, 195;
information, 189;
Internet, 7, 68, 73, 75, 95, 140, 186, 204.

accountability, 2, 36, 77–79, 89, 100, 151, 155, 173, 174, 191

ACNU. *See* Asociacion Cubana de las Naciones Unidas

ACTA. *See* Anti-Counterfeiting Trade Agreement

activism, 17, 201

actors:
global, 6, 195;
new, 24, 45–46, 48, 51–52, 54, 138, 207;
non-governmental, 21, 69, 74;
non-state, 2, 8, 13, 15, 47, 49–54, 56, 58, 109, 116, 121, 138, 163, 164, 167, 170;
private, 8, 39, 119, 167, 171, 181, 182, 192, 198, 206;

state, 2, 7, 49, 58, 87, 98, 168.

AfICTa. *See* Africa ICT Alliance

Africa ICT Alliance (AfICTa), 72

agonistic, 103, 105

ALAC. *See* At-Large Advisory Committee

algorithms, 6, 7, 146

ambassador, 10–11, 13, 21, 29, 32, 41, 111;
digital, 5, 10;
thematic, 10.

Amendment 138, 205

American Association for the Advancement of Science (AAAS), 10, 18, 112–13, 117

American Association of Independent Music (A2IM), 205

Anti-Counterfeiting Trade Agreement (ACTA), 155

APC. *See* Association for Progressive Communication (APC)

APIG. *See* Association for Proper Internet Governance (APIG)

arbitration, 33, 43, 147, 156

Argentina, 16, 28, 72, 185–87, 191–92

artificial intelligence, 6, 93, 107, 135, 138

Asociacion Cubana de las Naciones Unidas (ACNU), 72

Association for Progressive
 Communication (APC), 72, 77–79
Association for Proper Internet
 Governance (APIG), 72, 79
At-Large Advisory Committee (ICANN
 ALAC), 43
Australia, 28, 72, 136, 162, 168–69
Authority:
 adaptive, 96–100, 102–05;
 constituted, 95–100, 102–05.

Bilateral Investment Treaty (BIT), 146
BIT. *See* Bilateral Investment Treaty
Brazil, 16, 41, 72, 75, 121, 186–87, 191
Brazil, Russian Federation, India,
 China, and Republic of South Africa
 (BRICS), 68, 73, 75
BRICS. *See* Brazil, Russian Federation,
 India, China, and Republic of South
 Africa
business sector, 3, 6, 14, 53

CAS. *See* Copyright Alert System
ccTLD. *See* Country Code Top-Level
 Domain
centralization, 24, 35, 37
Centre for Internet and Society (CIS,
 India), 78
CETA. *See* Comprehensive Economic
 and Trade Agreement
CFSP. *See* Common Foreign and
 Security Policy
Charter of Fundamental Rights, 198
China, 32, 72, 76, 82, 96, 109–10,
 136, 139
CIS. *See* Centre for Internet and Society
CJEU. *See* Court of Justice of the
 European Union
climate change, 31, 51, 88, 93, 105,
 124, 148
CNCI. *See* Comprehensive National
 Cyber Security Initiative
CoE. *See* Council of Europe
Cold War, 9, 48, 108
Colombia, 16, 188, 189, 191–92

Common Foreign and Security Policy
 (EU CFSP), 131, 137
Common Security and Defense Policy
 (EU CSDP), 132
commons, 18, 41, 97, 158
communication:
 cross-cultural, 15, 107, 116–17;
 global, 195, 202.
community:
 academic, 46, 72, 77;
 epistemic, 35;
 Internet, 33, 35–37;
 technical, 52, 66, 83, 98, 114.
Comprehensive Economic and Trade
 Agreement (CETA), 143, 147, 158
Comprehensive National Cyber Security
 Initiative (US CNCI), 125
Conference on Global Internet
 Governance Actors, Regulations,
 Transactions and Strategies
 Conference (GIG-ARTS), 12, 18
consensus, 3, 36, 61, 71, 89, 109, 121,
 135, 170, 182
constitution, 30, 32, 37, 43, 152, 155,
 184, 188
constitutionalist, 72, 77–79, 81–83
content:
 access to, 101, 203;
 harmful, 17, 182;
 illegal, 17, 182.
Copyright:
 alert system, 205;
 control, 203;
 infringement, 182, 185, 203–04,
 206–07.
Copyright Alert System (CAS), 205
corporations, 8, 68, 78, 122, 156,
 196–97, 200–02
Council of Europe (CoE), 198
Country Code Top-Level Domain
 (ccTLD), 37
Court of Justice of the European Union
 (CJEU), 183, 184, 188, 201, 206
courts, 7, 100, 155, 156, 180–81, 183,
 186, 207

CSDP. *See* Common Security and Defense Policy
cybercrime, 17, 81, 83, 132–33, 138–39
cyberdefense, 127, 128, 132–33
cyberdiplomacy,130–31, 133–36
cybersecurity, 2, 6–7, 10–11, 13, 15, 79, 81–82, 97, 99, 107, 109–11, 114, 116–17, 119, 122–29, 131–40

DARPA. *See* Defense Advanced Research Projects Agency
data:
  big, 55, 138, 197, 206, 208;
  data protection, 6, 16–17, 123, 126, 138, 148, 152–153, 156, 192, 197–202, 208;
  economy, 150, 151;
  flow, 13, 15–16, 143–44, 148–51, 158;
  subjects, 150, 197, 199;
  transfer, 136, 198–201.
Data Protection Directive 1995 (EU DPD), 153, 198, 200
DEA. *See* Digital Economy Act 2010
decentralization, 21, 37
Defense Advanced Research Projects Agency (US DARPA), 114
definitional struggle, 62, 63, 71, 83
democracy, 5, 7–8, 11, 18, 22, 37, 40, 77–78, 90, 140, 144, 147, 149, 155
Denmark, 1, 10, 168
Department of Commerce (US DoC), 36, 43, 100, 112, 114, 126, 201–2
digital:
  companies, 1, 103;
  diplomacy, 2, 9, 11, 87–89, 93–94, 97–98, 101, 103–4;
  divide, 6, 40, 68–70, 74, 80, 84;
  environment, 2, 14, 87–89, 94–96, 99–100, 102, 104;
  issues, 1, 143;
  labor, 151, 156;
  natives, 203.
Digital Economy Act 2010 (UK DEA), 204

Digital Millennium Copyright Act (US DMCA), 179, 182, 185
digitalization, 5, 7, 179, 182, 185
diplomacy:
  actors, 15;
  agent, 8;
  classical, 13, 21, 24–25, 29–30, 33–34, 37–38, 40;
  confidential, 32;
  contemporary, 8–10;
  cultural, 107–8, 130;
  for science, 18, 107, 111–12;
  hyphenated, 30;
  multilateral, 24, 51;
  nature of, 8;
  new, 12, 15, 26, 32, 47, 51–57, 143;
  non-governmental, 30;
  parliamentary-style, 52;
  practice, 1–4, 10–15, 23, 31, 45, 50, 55, 57–58, 107;
  process, 9, 90;
  pseudo, 31;
  regalian, 10;
  science diplomacy, 9–12, 14–15, 18, 29–30, 85, 87–88, 90, 96–97, 99, 101–3, 105, 107–13, 115, 117, 119–20, 122–24, 126, 128–32, 135–40
  science for, 29, 90, 120;
  shuttle, 33, 42;
  terrain, 1, 11;
  traditional, 51, 111, 180, 181.
diplomatic culture, 25, 35, 248
disarmament, 17, 49
discourse:
  analysis, 14, 62, 66;
  coalitions, 14, 61, 63–65, 67, 71–73, 75, 78–80, 82, 83, 84;
  institutionalization, 63.
discursive order, 14, 62, 66–67, 69, 71–72, 80, 83
disinformation, 10, 27
disruption, 4–8, 56, 58, 126, 148, 163, 166, 173, 199, 205;
  democracy, 7;

digital, 4–5, 8;
economy, 7;
geopolitics, 7;
governance, 6.
disruptive technology, 55–56, 174
DMCA. *See* Digital Millennium
Copyright Act
DNS. *See* Domain Name System
DoC. *See* Department of Commerce
domain names, 52, 68, 100, 114
Domain Name System (DNS), 44, 52,
68, 73, 100–1
DPD. *See* Data Protection Directive
1995

EC. *See* European Commission
ECHR. *See* European Convention of
Human Rights
ECJ. *See* European Court of Justice
e-commerce, 28, 138, 182–83
Economic Commission for Latin
America and the Caribbean
(ELAC), 72
economic development, 4, 55, 80–81,
124, 167
EDPS. *See* European Data Protection
Supervisor
education, 18, 34, 55
EEAS. *See* European External Action
Service
ELAC. *See* Economic Commission for
Latin America and the Caribbean
ENISA. *See* European Union Agency for
Network and Information Security
EP. *See* European Parliament
ePrivacy Directive. *See* Privacy and
Electronic Communications Directive
ESCWA. *See* United Nation Economic
and Social Commission for Western
Asia
Estonia, 10, 183
ETSI. *See* European
Telecommunications Standards
Institute

EU-US Privacy Shield Agreement,
201–2, 206
EU. *See* European Union
European Commission (EC), 88, 99,
129–31, 134, 177
European Convention of Human Rights
(ECHR), 198, 205
European Court of Justice (ECJ), 16,
159, 193, 201
European Data Protection Supervisor
(EDPS), 208
European External Action Service
(EEAS), 135, 137, 140
European Parliament (EP), 140, 158–59,
169, 172–73, 203–5, 207–8
European Telecommunications
Standards Institute (ETSI), 162,
166–67, 172
European Union (EU):
citizen, 200–1, 206, 208;
member states, 3, 6, 12, 82, 111,
130–34, 137–38, 140, 161–64, 167,
169, 171, 173, 183, 198, 200, 205.
European Union Agency for Network
and Information Security (ENISA),
131, 134, 137

Facebook, 1, 11, 134, 183, 192, 201–2,
206, 208–9
fairness, 7, 33, 36, 42
fake news, 183
FANCV. *See* Fundación Argentina a las
Naciones Camino a la Verdad
FCC. *See* Federal Communications
Commission
Federal Communications Commission
(US FCC), 95
finance, 5, 7, 191
Finland, 10
foreign affairs, 2, 8–10, 55, 111,
121, 130
Foucault, Michel, 143, 157
fragmentation, 7–8, 43, 73, 77, 94, 138;
decentralized, 40, 68, 99, 107;
Internet, 94.

France, 165, 168, 203–4, 206, 208
free software, 24, 202
Free Trade Agreement (FTA), 13,
   15–16, 143–49, 151, 154–55
freedom:
   fundamental, 17, 177, 196–98;
   Internet, 94, 136, 139, 207.
freedom of expression, 17, 79, 88, 102,
   174, 177, 179–80, 182–83, 187–90,
   192, 207
FTA. *See* Free Trade Agreement
Fundación Argentina a las Naciones
   Camino a la Verdad (FANCV), 72

G20. *See* Group of Twenty
G77. *See* Group of Seventy-Seven
GAC. *See* Government Advisory
   Committee
GAFA. *See* Google, Apple, Facebook,
   and Amazon
Gates, Melinda, 3, 17
GATT. *See* General Agreement on
   Tariffs and Trade
GCIG. *See* Global Commission on
   Internet Governance
GCSC. *See* Global Commission on the
   Stability of Cyberspace
GDPR. *See* General Data Protection
   Regulation
General Agreement on Tariffs and Trade
   (GATT), 24, 28, 153
General Data Protection Regulation (EU
   GDPR), 153, 199, 202, 206
Generic Names Supporting Organization
   (ICANN GNSO), 36
Geneva Declaration of Principles, 69, 76
Geneva Plan of Actions, 69
GGE. *See* Group of Governmental
   Experts on Advancing responsible
   State behavior in cyberspace in the
   context of international security
GIG-ARTS. *See* Conference on
   Global Internet Governance Actors,
   Regulations, Transactions and
   Strategies Conference

Global Commission on Internet
   Governance (GCIG), 99
Global Commission on the Stability of
   Cyberspace (GCSC), 99
Global System for Mobile
   Communication (GSM), 166
globalization, 4–5, 7–8, 16, 31, 48–49,
   124, 145–46, 167, 179
GNSO. *See* Generic Names Supporting
   Organization
Google, 134, 183, 185–89, 191–93, 202
Google, Apple, Facebook, and Amazon
   (GAFA), 1, 5
governance:
   algorithmic, 6;
   global, 1–6, 8, 12, 32, 45, 47–50,
      56–59, 69, 97, 121;
   global Internet, 1, 4–5, 10–12, 17, 83,
      99, 109, 123, 189;
   self-, 68, 157;
   transnational, 1, 3–4.
Government Advisory Committee
   (ICANN GAC), 33
governmentality, 15–16, 143, 145, 157
GPD. *See* Global Partners Digital
graduated response, 17, 196, 203–9
Group of Governmental Experts on
   Advancing responsible State behavior
   in cyberspace in the context of
   international security (UN GGE),
   2–3, 17, 123
Group of Seventy-Seven (G77), 72,
   75–77, 84
Group of Twenty (G20), 98–99
GSM. *See* Global System for Mobile
   Communication

HADOPI. *See* Haute Autorité pour la
   diffusion des oeuvres et la protection
   des droits sur Internet
Haute Autorité pour la diffusion des
   oeuvres et la protection des droits sur
   Internet (French HADOPI), 204–7
heterarchy, 121, 125
hierarchy, 26, 40, 91, 138, 154, 156, 170

IACHR. *See* Inter-American
Commission on Human Rights
IANA. *See* Internet Assigned Numbers
Authority
ICANN. *See* Internet Corporation for
Assigned Names and Numbers
ICC. *See* International Chamber of
Commerce
ICRC. *See* International Committee of
the Red Cross
ICT. *See* Information and
Communication Technology
ideology, 26,197
IDP. *See* Internet Democracy Project
IEEE. *See* Institute of Electrical and
Electronic Engineers
IETF. *See* Internet Engineering Task
Force
IFLA. *See* International Federation of
Library Associations and Institutions
IFTA. *See* Independent Film and
Television Alliance
IGF. *See* Internet Governance Forum
IGO. *See* Intergovernmental
Organization
ILETS. *See* International Law
Enforcement Telecommunications
Seminar
ILO. *See* International Labor
Organization
IMF. *See* International Monetary Fund
IMO. *See* International Maritime
Organization
INAI. *See* Instituto Nacional de
Transparencia, Acceso a la
Información y Protección de Datos
Personales
inclusiveness, 34, 104
Independent Film and Television
Alliance (IFTA), 205
Information and Communication
Technology (ICT), 55
information society, 67, 69–71, 75,
77–78, 81, 82, 113, 133
innovation:

market, 5;
technical, 26;
technological, 23, 45, 54, 73, 87, 89,
91, 100–1, 166–67
Institute of Electrical and Electronic
Engineers (IEEE), 72–73
institution, 23, 35, 46, 58, 63, 70, 100,
135 ,147
institutionalization, 30, 50, 63, 71, 121,
125, 127, 131, 133, 135, 137, 143,
146–47, 162–63
Instituto Nacional de Transparencia,
Acceso a la Información y Protección
de Datos Personales (Mexico's INAI,
formerly IFAI), 188
Inter-American Commission on Human
Rights (IACHR), 190
intermediary, 4, 17, 184–85;
liability, 17, 187, 191;
private, 7, 37, 40;
Intergovernmental Organization (IGO),
7, 24, 35–37, 68, 73, 83, 203
intergovernmental, 7, 24, 30–32, 35–37,
53–54, 68, 71, 73–75, 77, 83, 103,
132, 136, 190, 196
intermediation, 13, 21, 24, 30, 32, 34,
37, 40, 247
international:
agreements, 2, 32, 124, 130;
cooperation, 4, 6, 16, 43, 48, 51, 88,
107, 128, 137, 167–68, 170;
development, 2, 7, 15;
development aid, 7, 9;
human rights law, 78–79, 83;
law, 37, 68, 76, 81, 131, 134, 168;
trade, 144, 182.
International Chamber of Commerce
(ICC), 72, 74, 213
International Committee of the Red
Cross (ICRC), 39
International Federation of Library
Associations and Institutions
(IFLA), 72
International Labor Organization
(ILO), 39

International Law Enforcement
Telecommunications Seminar
(ILETS), 168–69
International Maritime Organization
(IMO), 39
International Monetary Fund (IMF), 146
International Organization (IO), 3–4,
33–34, 52, 66, 68–70, 72, 79, 83,
98, 109–11, 113, 115–16, 122–23,
131, 179
International Telecommunications Union
(ITU), 67–68, 162
Internet Assigned Numbers Authority
(IANA), 35–36, 40, 42–44, 53, 100
Internet Corporation for Assigned
Names and Numbers (ICANN), 24,
32–40, 42–43, 53, 68, 71–72, 74, 76,
79, 97, 100–2, 130
Internet Democracy Project (IDP),
72, 78
Internet diplomacy, 2–5, 8, 10–12, 108;
research agenda, 8, 10, 11.
Internet Engineering Task Force (IETF),
35, 101, 110
Internet Governance Forum (IGF), 6,
18, 30–32, 35, 40–41, 53–55, 70, 97,
102, 114–115, 130, 140
Internet infrastructure, 2, 14, 77,
110, 138;
critical, 77, 125–26, 129, 132,
135, 138.
Internet Service Provider (ISP), 181,
203–5
Internet Society (ISOC), 54, 72
interoperability, 33
IO. *See* International Organization
ISOC. *See* Internet Society
isomorphism, 114, 116
ISP. *See* Internet Service Provider
ITU. *See* International
Telecommunications Union
IUR. *See* International User
Requirements

Japan, 25, 72, 132, 136

Japan Business Federation (JBF), 72
Japan Information Technology Services
Industry Association (JISA), 72
Japan Network Information Center
(JPNIC), 72
Japan Registry Services (JPRS), 72
JBF. *See* Japan Business Federation
JISA. *See* Japan Information Technology
Services Industry Association
JPNIC. *See* Japan Network Information
Center
JPRS. *See* Japan Registry Services
jurisdiction, 6, 55, 97

KLRCA. *See* Kuala Lumpur Regional
Centre for Arbitration
Klynge, Casper, 10, 11
knowledge, 4, 9, 18, 22, 55, 72, 92–93,
95, 98, 107–8, 110–12, 114, 116,
122, 155, 166, 181, 183, 185–86,
197, 200
Korea, 25, 28, 72, 203
Kuala Lumpur Regional Centre for
Arbitration (KLRCA), 43

labor, 5, 7, 31, 151, 156, 158, 168
Law Enforcement Agency (LEA), 161,
163–64, 169–71, 173–74
Law Enforcement Monitoring Facility
(LEMF), 161–62, 172, 174
lawful interception, 16, 161–65, 167–70,
172–73, 175
LEA. *See* Law Enforcement Agency
legitimacy, 7, 22, 28, 30, 35, 53, 77–78,
83, 90, 98, 101, 181
LEMF. *See* Law Enforcement
Monitoring Facility
lobbying, 87, 98, 143, 156, 206

Ma, Jack, 3
Malaysia Network Information Centre
(MYNIC), 43
mediation, 9, 25, 37, 43;
cultural, 9;
intercultural, 9.

meta organization, 34–35, 37, 42
Mexico, 16, 28, 72, 187–88, 191
Mexico, Indonesia, Republic of Korea, Turkey, and Australia (MIKTA), 28
Microsoft, 11, 123, 134, 202
Middle East, 25
MIKTA. *See* Mexico, Indonesia, Republic of Korea, Turkey, and Australia
Motion Picture Association of America (MPAA), 205
MPAA. *See* Motion Picture Association of America
MS. *See* European Union Member State
multilateralism, 8, 21, 25, 33–34, 37, 40, 121, 128, 147
multistakeholder, 3, 6, 10, 12, 13, 15, 42–43, 53–54, 57, 65, 69, 71–72, 76–77, 82–83, 88–89, 97–104, 108, 114–16, 121, 123, 136, 139, 177, 191, 192;
  approach, 3, 98, 115, 123;
  participation, 3, 6.
multistakeholderism, 2, 8, 14, 35, 37, 47, 51–54, 57, 67, 69, 70, 71, 82, 98–99, 109, 115–16
mutation, 3–5, 7–9, 11, 146
mutual adjustments, 65, 72, 80
MYNIC. *See* Malaysia Network Information Centre

Narrative Policy Framework (NPF), 64
nation, 1, 7–9, 57, 67, 76, 109, 110–11, 115, 122, 152, 202
National Research Council (NRC), 113, 124
National Security Agency (US NSA), 127, 133, 139, 172
National Telecommunications & Information Administration (US NTIA), 100
neoliberal, 14, 16, 68, 70, 72–74, 81–84, 103, 145–46, 148, 154, 157–58
Net Neutrality. *See* network neutrality

NetMundial. *See* NetMundial Initiative—Global Multistakeholder Meeting on the Future of Internet Governance
NetMundial Initiative—Global Multistakeholder Meeting on the Future of Internet Governance, 53–54
Network and Information Security (NIS), 132, 134–35, 140
network neutrality, 7, 16, 95–96
NGO. *See* Non Governmental Organization
NIS. *See* Network and Information Security
Non Governmental Organization (NGO), 22, 30, 32, 34, 77, 109, 111–12, 122–23, 159, 180–81, 188
norm:
  behavior, 28, 35, 131;
  diffusion, 178–80, 185, 190;
  normative framework, 8, 179;
  normative value, 48.
NPF. *See* Narrative Policy Framework
NRC. *See* National Research Council
NSA. *See* National Security Agency
NTIA. *See* National Telecommunications & Information Administration

Obama Administration, 125–27, 139
OECD. *See* Organization for Economic Co-operation and Development
OEWG. *See* Open Ended Working Group
Office of Management and Budget (US OMB), 127
Office of the Science and Technology Adviser to the Secretary of State (US STAS), 124
OMB. *See* Office of Management and Budget
Ombudsman, 33, 42
Open Ended Working Group (UN OEWG), 3, 17, 40, 43
open source, 29, 204

openness, 5, 90, 96–97, 104, 135
Organization for Economic Co-operation
and Development (OECD), 122, 138,
152, 200, 206
Ottoman Empire, 25

P2P. *See* peer-to-peer
Parliament, 7, 25, 134, 137
PASA. *See* Peoples Advancement Social
Association
peer-to-peer (P2P), 203, 205;
file sharing, 203;
infringement, 205;
technologies, 203.
Peloponnesian League, 24
Peoples Advancement Social
Association (PASA), 72
Persia, 25
PICDRP. *See* Public Interest
Commitment Dispute Resolution
Procedure
pirate party, 203
platform, 6, 10, 14, 87, 104, 115, 123,
132–33, 144, 149–51, 156, 158, 181,
183, 191, 202;
communication, 202;
power, 87;
regulation, 6;
state, 10.
platformization, 7
policy:
dialogue, 53, 70;
diffusion, 13, 16–17, 178, 195–96,
201–3;
digital, 3, 49;
domain, 61, 63–65, 67, 71, 111, 124;
foreign, 4–5, 17–18, 48, 90, 107, 110,
112, 119–20, 122, 124–27, 135, 140,
180, 201, 207;
framework, 2, 4, 64, 133, 165,
168, 214;
principles, 17, 196;
problem, 62–63, 65–66, 69, 73, 75, 78,
80–81, 84;

public, 5, 7, 37, 70–71, 73, 76–77, 101,
136, 156, 191–92;
regime, 195–97, 201;
reverse diffusion, 201;
solution, 64–66, 69–70, 73, 75–76, 78,
80–81, 83–84;
space, 109, 111–15, 117;
statements, 64–65;
transfer trajectory, 206.
policymakers, 6, 108, 191, 203
policymaking, 61, 63, 70, 79, 101, 122,
128, 135, 171, 195, 199, 200, 201
political participation, 195, 270
politicization, 70, 94, 95
polylateralism, 8, 11
Postal, telegraph, and telephone agency
(PTT), 164
Post-Snowden Era, 197
power:
asymmetry, 197;
balance of, 22, 36;
centralized, 35;
imbalance, 54, 197;
market, 95;
mechanism, 54;
politics, 47, 137, 202, 207;
relations, 2, 5, 6, 91, 94, 104, 121;
relationships, 52, 63, 89, 92;
soft, 9, 11, 89, 104, 108, 120, 126,
132–33, 180, 209;
structures, 47.
privacy, 13, 16–17, 32, 78, 88, 90, 95,
102, 123, 126–27, 133, 136, 138–40,
144, 148, 150–54, 165–66, 170–71,
174–75, 177, 179–80, 182, 184,
188–89, 191–92, 195–202, 205–9;
consent-based policies, 202, 206;
consent-driven policies, 201;
right to, 78, 165, 171, 175, 180,
196–97, 208;
standard, 199–200, 202, 206.
Privacy and Electronic Communications
Directive (EU ePrivacy Directive),
199, 205
Privacy Shield, 126, 201–2, 206, 208

profiling, 7, 149, 198
protocol, 2, 4, 7, 14, 23, 37–38, 68, 100,
    123, 133
PTI. *See* Public Technical Identifiers
PTT. *See* Postal, telegraph, and
    telephone agency
public diplomacy, 9–11, 13, 15, 17–18,
    40, 107–12, 117, 180;
    2.0., 11.
Public Interest Commitment Dispute
    Resolution Procedure (PICDRP), 43
Public Technical Identifiers (ICANN
    PTI, IANA's successor), 43
public-private partnership, 171

Quasi (Autonomous) Non Governmental
    Organization (QUANGO), 37

Recording Industry Association of
    America (RIAA), 205
Regional Internet Registries (RIRs),
    36, 72
regulation, 6–7, 13–14, 17, 30, 38, 52,
    69, 74–75, 88, 132, 149–54, 157,
    159, 177–78, 181–84, 189, 192, 195,
    197, 199, 206–8
Response Policy Zone (RPZ), 100
revolving doors, 11
RIAA. *See* Recording Industry
    Association of America
rights:
    citizens', 77, 196, 206, 208;
    civil, 172, 196;
    consumer, 4;
    digital, 149, 159, 205;
    human, 4–7, 16, 55, 68, 78–79, 81, 83,
        88, 97, 100, 123, 131, 139–40, 145,
        159, 162, 165, 170, 172–74, 177,
        183, 186, 190, 196, 198;
    intellectual property, 6, 1,7 150.
Right to be forgotten (RTBF), 13, 16,
    177–79, 181, 185–93
Rio Earth Summit, 51
RIRs. *See* Regional Internet Registries
    (RIRs)

RPZ. *See* Response Policy Zone
RTBF. *See* Right to be forgotten
rule of law, 5, 7, 79, 89, 140, 172, 180

Safe Harbor Privacy Principles, 201
Samuelsen, Anders, 1
Schrems, Maximilian, 159, 201, 207
Science Technology and Innovation
    (STI), 107, 124
SDGs. *See* Sustainable Development
    Goals
security, 1–3, 5, 7–9, 49, 51, 55, 75–77,
    81–82, 88, 90, 96, 99, 107, 112,
    114, 117, 121–23, 125–28, 130–40,
    150, 152, 154–55, 161–72, 174–75,
    195, 200;
    national, 49, 51, 96, 112, 126–27,
        133, 136, 139, 155, 161, 163, 165,
        170, 171.
self-determination, 168, 197, 198
self-regulation, 52, 74, 182–83
sharing economy, 7, 151
Snowden, Edward, 139, 177, 190, 192
social:
    cohesion, 4–5;
    constructivist, 92–93;
    justice, 5;
    media, 6, 55, 94, 102, 108, 117, 130,
        138, 202;
    movements, 203, 208;
    networks, 7, 9, 25, 32, 34, 151, 181;
    transformation, 5.
Society for the Social Studies of Science
    (4S), 72
socio-legal studies, 178, 180
socio-technical controversy, 87, 89, 94
soft law, 7, 34
sovereignty, 4–5, 13, 16, 25, 30, 37, 44,
    46–47, 49, 76, 82, 94, 122, 147, 150,
    161, 182;
    digital, 5–6.
Spanish Government, 109
stability, 2, 8, 35, 99–100, 123
stakeholder, 8, 12, 30, 32, 66, 72, 83, 87,
    92–93, 97, 99, 101, 115, 116–17

STAS. *See* Office of the Science and Technology Adviser to the Secretary of State
stewardship, 30, 32, 36, 40, 177
STI. *See* Science Technology and Innovation
Sumerian League, 24
supranational, 135, 146–48, 200–1, 207
surveillance, 6–7, 13, 16–17, 78–79, 83, 87, 133, 140, 144, 150, 161–63, 166, 168, 169–75, 207;
 electronic, 6;
 state, 17, 87.
sustainability, 5, 51
Sustainable Development Goals (UN SDG), 51

taxation, 5, 7, 73–74, 136
technical management, 75, 77, 81
technical organization, 12, 113
techno-diplomacy, 39
techplomacy, 11
telecoms reform package, 205, 207
telegrams, 27, 32
Three-Strikes Policy, 203
Thucydides' Melian Dialogue, 21
TiSA. *See* Trade in Services Agreement
trade, 5–7, 13, 15–16, 22, 24, 28, 41, 96–97, 113, 143–49, 151–59, 179, 182, 200;
 digital, 28;
 international, 144, 182.
Trade in Services Agreement (TiSA), 15–16; 143–46, 148–49, 151–56, 158–59
Transatlantic Trade and Investment Partnership (TTIP), 143, 146–47, 159
transnational, 1–4, 6, 13, 16, 32, 43, 44, 48, 68, 144, 147, 154, 156, 159, 163, 167, 168, 180;
 power relations, 2, 5–6 52, 63, 89, 91–92, 94, 104;
 ruling, 32, 68.
transparency, 5, 7, 23, 49, 77–78, 90, 96–97, 99, 174, 191

Treaty organization, 33, 39
Trump, Donald, 27, 95, 100, 125, 127–28, 146, 158
TTIP. *See* Transatlantic Trade and Investment Partnership (TTIP)
Tunis Agenda For Information Society, 69
Tunis Commitment, 69
Twitter, 172, 202, 206, 208, 209

uberisation, 7
UK. *See* United Kingdom
UN High-Level Panel On Digital Cooperation, 3
UN. *See* United Nations
UNCTAD. *See* United Nations Conference on Trade and Development
undiplomatic, 13, 21, 23, 25, 27
UNESCO. *See* United Nations Educational, Scientific and Cultural Organization
UNGA. *See* United Nations General Assembly
United Kingdom (UK), 10–11, 25, 37, 83, 90, 159, 165, 168–69, 206, 208
United Nation Economic and Social Commission for Western Asia (ESCWA), 72
United Nations (UN), 3–4, 6, 12, 14, 76–77, 81, 84, 97, 99, 114–15
United Nations Conference on Trade and Development (UNCTAD), 24, 145
United Nations Educational, Scientific and Cultural Organization (UNESCO), 72, 77, 88, 94, 97, 111
United Nations General Assembly (UNGA), 3, 54
United States (US), 9, 15, 43, 68, 72–74, 90, 100, 107–11, 114–16, 124–26, 129, 132, 136, 139, 158–59, 162–63, 168, 172, 177, 184, 201
United States Agency for International Development (USAID), 9

United States Government (USG), 100, 172

US. *See* United States

USAID. *See* United States Agency for International Development

USG. *See* United States Government

US State Department, 27, 113, 125, 128

values, 2, 4–7, 13, 15, 26, 32–33, 39, 51, 89–90, 92, 94, 96, 98, 100, 104–5, 128, 130, 134, 136, 139, 140, 148–49, 152–53, 177–78, 180, 189–90, 192, 201, 206–8; cultural, 4; democratic, 4–5, 7, 15, 130, 134.

VCDR. *See* Vienna Convention on Diplomatic Relations

Verdier, Henri, 10

Vienna Convention on Diplomatic Relations (VCDR), 50

WEF. *See* World Economic Forum

Westphalia: 1648, 27; westphalian, 50, 68, 103, 169.

WGIG. *See* Working Group on Internet Governance

WIPO. *See* World Intellectual Property Organization

WMO. *See* World Meteorological Organization

Working Group on Internet Governance (WGIG), 4, 53, 69

World Economic Forum (WEF), 8, 97–98

World Intellectual Property Organization (WIPO), 24, 27, 33, 39, 43, 72, 97

World Meteorological Organization (WMO), 39

world order, 103, 120

World Summit on Information Society (UN WSIS), 2, 6, 17, 36, 40–41, 46, 51, 53, 55, 57, 62, 66, 68–74, 80–82, 97, 102, 130, 139

World Summit on Information Society +10 review process (WSIS+10), 12, 14, 62, 66, 67, 71–72, 74, 80, 82–84

World Trade Organization (WTO), 24, 28, 41, 97, 144, 146–47, 159

WSIS. *See* World Summit on Information Society

WSIS+10. See World Summit on Information Society +10 review process

WTO. *See* World Trade Organization (WTO)

zero-rating, 7

About the Editors and Contributors

Francesco Amoretti
Francesco Amoretti is professor of political science and of administration science at the University of Salerno. He is a member of the directive committee and of the editorial board of the *Political Communication* review, and founding co-editor and member of the editorial board of *Soft Power*, a Euro-American journal of historical and theoretical studies of politics. He is vice dean of the Italian Political Science Association (SISP), and director of the Internet and Communication Policy Center (ICPC). His most recent publications on "Future Directions in Global Political Economy: The Policy Strategy of International Monetary Fund in the Covid-19 Pandemic" appeared in *Partecipazione e Conflitto* 14(1), 2021 (with D. Giannone and A. Cozzolino).

Andrea Calderaro
Andrea Calderaro is senior lecturer in international relations, director of the Centre for Internet and Global Politics at Cardiff University, and Robert Schuman center for advanced studies fellow at the European University Institute. His research centers on transnational governance of technology, with a focus on cybersecurity and cyber diplomacy. He has conducted research and supported cyber capacity-building initiatives in Africa, Asia, The Middle East, Central America, and in EU institutions, and he serves as a member of the research committee of the Global Forum of Cyber Expertise (GFCE). He has been a visiting fellow at the California Institute of Technology (CalTech), Humboldt University, LUISS Guido Carli, University La Sapienza, and University of Oslo. He holds his PhD in Social and Political Sciences from the European University Institute.

Jean-Marie Chenou

Jean-Marie Chenou is associate professor in international relations at the department of political science and a member of the global studies research group at the University of Los Andes, Bogotá, Colombia. He holds a PhD in Political Science from the University of Lausanne, Switzerland. His research interests include global political economy, Internet governance, and the digital transformation in Latin America.

Domenico Fracchiolla

Domenico Fracchiolla is adjunct professor and research fellow of international relations at the University of Salerno. He is an adjunct professor of sociology of international relations at the LUISS University, Rome, an adjunct professor of international governance at Link Campus University, Rome, and an adjunct professor of history of international relations at Mercatorum University, Rome. He has been a visiting scholar at the UC Berkeley, California, USA; and at the SAIS, Johns Hopkins University, Washington DC, USA. He is member of the Internet and Communication Policy Center (ICPC), University of Salerno. He has been deputy director of the research center LUISS-LAPS, LUISS University.

Maria Francesca De Tullio

Maria Francesca De Tullio is PhD in "Human Rights. Theory, History and Practice," with a thesis in constitutional law, at the University of Naples Federico II, and with a research stay at the Université Paris 2 Panthéon Assas, CERSA, funded by the erasmus+ program of the European Union. She authored a monography concerning substantial equality and new dimensions of political participation. She worked as a postdoc at the University of Antwerp within the project cultural and creative spaces and cities, funded by the Creative Europe program of the European Union. Her main research areas are political representation and participatory democracy, counterterrorism and legal states of emergency, communication surveillance, competition law on the Internet, and the collective dimension of privacy in the era of big data.

Katharina E. Höne

Katharina E. Höne is director of research at the not-for-profit organisation DiploFoundation (diplomacy.edu). Her research explores the interplay between digital technology and diplomacy and the impact of digital technology and digitalization on international relations. In her recent work she looked at the challenges and opportunities of artificial intelligence for the conduct of diplomacy, the shifts in diplomatic practice towards digital tools in light of COVID-19, and the emergence of digital foreign policy.

Nanette S. Levinson

Nanette S. Levinson, the School of International Service, American University, Washington, DC, is one of three faculty directors of the Internet Governance Lab. She is a past chair of the Global Internet Governance Academic Network (GigaNet). A co-editor of *Researching Internet Governance: Methods, Frameworks, Futures* (MIT Press, 2020), her research and teaching focus on knowledge transfer in complex, cross-national systems; internet governance; science and technology policy; cross-cultural communication; and interorganizational learning and innovating in a range of settings. She received her bachelor's, master's, and doctoral degrees from Harvard University.

Robin Mansell

Robin Mansell holds a chair in new media and the Internet in the department of media and communication, London School of Economics and Political Science. Past president of the international association of media and communication research and member of the board of directors of TPRC (Research Conference on Communications, Information and Internet Policy), she is author of numerous scholarly papers and books including *Imagining the Internet: Communication, Innovation and Governance* (Oxford 2012) and *Advanced Introduction to Platform Economics* (with W. E. Steinmueller, Edward Elgar 2020).

Meryem Marzouki

Meryem Marzouki is a senior academic researcher in political sciences with the CNRS. She teaches Internet regulation and governance at Sorbonne Université. Her research focuses on global internet governance actors, issues, and institutionalization processes, as well as public and private digital regulations, exploring the resulting transformations of human rights, democracy, and the rule of law. She is the French management committee member of the EU COST Action GDHRNet (Global Digital Human Rights Network). She initiated in 2017 the multidisciplinary network and conference series on 'Global Internet Governance—Actors, Regulations, Transactions and Strategies' (GIG-ARTS). A co-founder of many digital rights organization and coalitions since 1996, she also participated as civil society stakeholder and served as independent expert in international institutions processes (UN WSIS 2002–2005, UN IGF 2006–2015, Council of Europe 2005–2013, OECD 2008–2015).

Giuseppe Micciarelli
Giuseppe Micciarelli is a political philosopher and legal sociologist (University of Salerno). He is the author of numerous essays and articles on the themes of democracy in transformation, institutional theory, global governance and social movements. In July 2019, he was awarded the Elinor Ostrom Award, the most prestigious award in the field of the defence and management of the commons. With his research he supports numerous groups, networks, associations, and local administrations for the creation and implementation of new participatory institutions. Currently, he is president of the Observatory on the Commons, participatory democracy and fundamental rights of the city of Naples. His latest book is titled *Commoning. Urban Commons and New Institutions. Materials for a Theory of Self Organisation* (second edition, 2021).

Nicola Palladino
Nicola Palladino (PhD in Sociology, Social Analysis, and Public Policy) is research fellow at the school of law and government of the Dublin City University. He is also a member of the digital constitutionalism network supported by the Center of Advanced Internet Studies in Bochum, Germany, and of the Internet and Communication Policy Center at the University of Salerno. His main research interests include global Internet governance processes, digital constitutionalism, platform governance, content governance, AI ethics, and regulation.

Claire Peters
Claire Peters's research is broadly concerned with the interaction between political and technical considerations in the design of Internet and data policy. As a MSc in international relations at the University of Bristol, her dissertation drew upon STS theory to examine Internet governance as a contextualized form of technology governance. She currently lives in San Francisco.

Krisztina Rozgonyi
Krisztina Rozgonyi is assistant professor at the department of communication of the University of Vienna, where she is also involved with managing a platform to bridge research with policy agendas. Her research focuses on specific aspects of media governance, the governance of spectrum and copyright, and the representation of public interest, democratic values, and fundamental rights within complex and highly 'technocratized' policy and regulatory processes. Dr Rozgonyi was a senior regulator and worked with international and European organizations, national governments, and regulators as adviser on media freedom, spectrum policy, and digital platform governance.

Mauro Santaniello

Mauro Santaniello is researcher and adjunct professor at the department of political and social studies of the University of Salerno, where he teaches Internet governance and digital policy. He is a founding member of the "Digital Constitutionalism Network" (DCN), co-founder and vice director of the "Internet & Communication Policy Centre" (ICPC), and a member of the international research network on "Global Internet Governance Actors, Regulations, Transactions and Strategies" (GIG-ARTS), and of the "Global Internet Governance Academic Network" (GigaNet). His research activity covers areas of global governance such as cybersecurity, standard-setting, platform regulation, digital rights, and powers.

Katharine Sarikakis

Katharine Sarikakis is professor of media governance at the department of communication, University of Vienna. She is the director of the media governance research lab and the Jean Monnet Centre of Excellence FREuDe and research associate at the University of Witswatersrand, South Africa. In recent years Katharine held the Jean Monnet chair and the Santander chair of excellence (Carlos III Universidad Madrid). She was visiting senior fellow at the London School of Economics and Political Science and received the European award in excellence in teaching. Katharine consults regularly with international organisations, such as CoE, FRA, and OSCE. She has served in leading positions the scholarly community at ECREA; ICA, IAMCR since 1998 and was the youngest-ever elected vice president of IAMCR.

Yves Schemeil

Yves Schemeil is emeritus professor of global and comparative politics, Sciences Po Grenoble, and visiting professor in geopolitics, Grenoble Ecole de Management (GEM). He has taught and lectured about international affairs in several universities, among which UCLA, Berkeley, Chicago, Geneva, and Tokyo. Published in five languages, his books and articles contribute to the academic debate about global organizations and world order. He is currently working with several teams of researchers from various countries on Internet governance, interorganizational networks, multilateralism, rising powers, and intercultural relations.

www.ingramcontent.com/pod-product-compliance
Lightning Source LLC
Chambersburg PA
CBHW031352290326
41932CB00044B/986